On Mathematics, Unity, and the Good

On Mathematics, Unity, and the Good

Plato, Whitehead, Aristotle

David A. White

BLOOMSBURY ACADEMIC
NEW YORK • LONDON • OXFORD • NEW DELHI • SYDNEY

BLOOMSBURY ACADEMIC
Bloomsbury Publishing Inc, 1359 Broadway, New York, NY 10018, USA
Bloomsbury Publishing Plc, 50 Bedford Square, London, WC1B 3DP, UK
Bloomsbury Publishing Ireland, 29 Earlsfort Terrace, Dublin 2, D02 AY28, Ireland

First published in the United States of America 2026

A catalog record for this book is available from the Library of Congress.

ISBN: HB: 978-1-6669-6331-1
ePDF: 979-8-216-25813-1
eBook: 978-1-9787-6556-6

Typeset by Deanta Global Publishing Services, Chennai, India

The qualities of number appear to lead toward truth. (Republic 7, 525b)

A puzzle, and source of criticism for those lacking a solution, is how mathematical elements and principles are related to the good and the beautiful. (Metaphysics 14, 1091a30-32)

To my students who, for many years, made me think about metaphysics much more than I might have been inclined to do on my own.

Contents

Preface

The title of this work may occasion surprise. A connection established via unity between mathematics and the good? Prompt assurance: for the three philosophers cited in the title, precisely this connection was a subject of speculative and critical analysis. The work to follow explores a series of related themes as this connection emerges in Plato, Whitehead, and Aristotle. The primary context of inquiry falls into the branch of philosophy now identified as metaphysics. Our principal burden is to show how developments in mathematics and mathematical understanding can indicate, illustrate, and contribute to assessing metaphysical considerations when accounts of reality are approached from the standpoint of an articulated representation of the good.

The work's argument situates the good in ways endowing such a reality—or perhaps for now, and in deference to the philosophically skeptical regarding the precarious status of "the good," putative reality—with substantial, determinate content. The work will also provide mechanisms and particular illustrations for critically evaluating metaphysical claims derived from that content. The narrative strategy consists of a gradual unfolding of abstract contexts characterized by mergers of mathematics—ancient as well as modern—and metaphysics. This progression leads directly into a formulation of the good (with the Platonic notion of the good standing as a model) pitched at a level of high generality. Then, with the good in view, the argument turns to discussions of metaphysical concerns influenced by and contoured according to mathematical realities in senses to be stated and justified.

Plato and Aristotle were students of mathematics but were primarily important as contributors to the philosophy of mathematics; Alfred North Whitehead made epochal contributions to both disciplines. The work traces themes involving all three thinkers as their approaches to reality converge on the intersections of broadly based metaphysical inquiry and mathematical concerns and discoveries. In terms of technique and strategy, the work straddles a neat procedural separation between traditional scholarship and controlled speculation. Discussion will indicate as forthrightly as possible interpretive exercises that move, sometimes boldly, from the former domain and its tight rules of procedure to the latter arena endowed as it is with more loosely defined limits of acceptable discourse. It may be emphasized at the outset that the speculative development pursued does not stand as reckless manipulation of texts, advancing a novel reading just for the sake of novelty itself. The work to follow is a trial, an argued attempt at saying something useful about metaphysical concerns by means of connections between mathematics and the good.

The work is in two parts. Part I opens with a discussion in Chapter 1 of the *Phaedo* on the generation of numbers, thus directly introducing Platonic metaphysics into a mathematical setting. Chapter 2 recalls Whitehead's positive reference (in *Process and Reality*) to Plato's use of triangles in the *Timaeus* and to the historical significance of this mathematical element in cosmological speculation; thus, the historical framework is significantly broadened, from antiquity to a locus of inquiry much closer to our own time. We also transfer the import of Whitehead's reading of the prescient Platonic use of triangles for metaphysical purposes into the domain of numbers when they are construed, as the *Phaedo* and other dialogues assert, according to certain fundamental metaphysical positions.

In Chapter 3, Platonic metaphysics in a post-*Phaedo* format serves as the backdrop for inquiry; here we clarify the structure and progression of grouping as an integral element in collection and division, a methodology Plato advances for metaphysical purposes in several later dialogues. Mathematics proper now moves in tandem with metaphysics, given that approaches to grouping are also important to various areas in modern mathematics. As a result, Chapter 4 introduces a set of related general terms linking the Platonic approach to collection and division—collection seen as a type of grouping—with the structure of an Abelian group, first formulated by mathematicians in the late eighteenth century and a foundational component in group theory. In addition, Chapter 4 shows how structural considerations that define Abelian groups, duly integrated with the basic elements of collection and division, establish a theoretical framework adaptable for evaluating a variety of diversified positions in metaphysical thought.

The remainder of the work develops this mathematically oriented approach to metaphysical assessment.

Part II opens with Chapter 5, devoted to examining selected arguments combining finely grained analytical criticism and speculative possibility as developed in Aristotle's extensive (and, according to Aristotle translator and commentator Julia Annas, "ruthless") critique of Platonic philosophy of mathematics (as stated in *Metaphysics*, Books M and N). Appeals to various approaches to unity, and the need to distinguish these approaches with care, inform many of Aristotle's critical observations.

Chapter 6 links the Aristotelian criticisms of pivotal positions in Platonic mathematical thinking with parallel concerns in the later Platonic methodology of collection and division. Here the interplay between the notions introduced to criticize mathematical elements (in Chapter 5) and the same notions devoted to evaluating the metaphysical positions considered in Chapter 6 suggests an intriguing fundamental identity. Can such sameness, when functioning at a high level of generality, be exploited in purely metaphysical settings?

Tradition has it that late in life, Plato delivered a lecture on the good, which the audience found puzzling since its content, according to those attending, concentrated on mathematics. Chapter 7 combines aspects of this circumstance with certain features of Whitehead's late essay, "Mathematics and the Good," thereby establishing a provisional foundation for integrating a version of the good with directed mathematical investigations. At this junction, Plato returns. The account of the good in the *Republic*, in conjunction with the intimacy that the famous (or, for some, infamous) divided line confers on mathematics in relation to Platonic Forms, generates a framework suitable for the investigation of metaphysical concerns. Review of these seminal positions leads the primary argument of Chapter 7 to an initial separation of mathematics from metaphysics in order to identify structural samenesses—called "commonalities"—traversing both disciplines. Analysis of the backdrop array of interlocking Aristotelian criticisms of Plato on mathematics (as reviewed in Chapter 5 and applied in Chapter 6) then generates a series of counterpart questions, concerns, and problems of broad theoretical relevance.

Although the emergent commonalities arise from an arena of inquiry limited to a specific area in mathematics—Abelian groups as originary group theory—this foundational study serves as a point of departure to the most speculative phase of the study. Chapter 7 introduces two seminal descriptions of the good in Plato—holding all things together (*Phaedo* 99c) and the apparently unique dependency of the Forms on the good (*Republic* 509b)—which support inquiry on a more inclusive scale than that discussed so far. In fact, implications from the reflections stated in Chapter 7 point to and justify a significantly expanded universe of metaphysical inquiry and assessment based on mathematical foundations established in conjunction with an articulated notion of the good. Aristotle's guarded, ultimately skeptical comments in the *Metaphysics* on the relation between mathematics and the good provide procedural guidance as well as sobering control on both the articulation and limits of the good in this regard.

Chapter 8 concludes the work. How could the good be structured in modes accessible to human experience and articulation, especially in abstract regions of metaphysical speculation filtered through mathematical elements? Chapter 8 opens by introducing a systematic interplay of "meta-principles" and "meta-concepts"—notions of high generality: for example, class, unity, linkage (or relation), and distinctions between and among dimensions of difference. When we bring the good down to earth, so to speak, according to these parameters and allow them to enjoy various modes of applicability, systematic metaphysics executed after Plato can exhibit accessible features of the good receptive to being extricated, identified, deployed, and, if deemed necessary, critically evaluated. Thus, if the good as characterized here has made its presence felt at certain points throughout the history of metaphysics (as it manifestly has in the speculative thought of Whitehead), we may justifiably conclude that any instance of inquiry drawn from past (and, indeed, ongoing) dimensions of metaphysical reflection can be investigated within the deployment and carefully articulated structure of the good as the argument herein has portrayed this structure.

Is there a "best" metaphysics? If not (and the reasons to support this denial would surely be of interest), can a given metaphysics become "better" than it was when originally posited? Implementation of this possibility rests in turn on exploring a perspective on comparative metaphysics, a procedural ploy discussed in Chapter 8 in the context of pursuing argumentatively—from both substantive and heuristic perspectives—the meta-principles and meta-concepts introduced in this chapter as the work's speculative culmination. The practical matter of how to apply these conclusions also includes reflection regarding the crucial importance of the imagination (Whitehead returns in this context with a brief but discerning contribution) in "doing" this kind of diversified metaphysics, especially within parameters established in part by mathematical elements. Also included in this vein—thoughts suggesting that putative practitioners of this program have a lively sense of speculative adventure throughout the hard work of thinking as extensively and systematically as possible with the inherently abstract and historically varied vocabulary which the work recommends.

The study as a whole is thoroughly rooted in selected texts of the three philosophers cited in the title; however, for the purpose of reducing possible misconceptions, we emphasize here in the Preface that the overall argument is entirely the author's speculative and admittedly ambitious development of positions advanced by an integrated reading of these epochal thinkers. The reader's discretion will determine whether this argument, especially in its outer reaches, can stand by itself as cogent and compelling. We may note here only that the feasibility of such reflection—especially construing the pivotal position of the good as itself historically conditioned and also grounding individual instances of metaphysics appearing throughout the history of metaphysics—will depend on the persuasiveness of the reasons leading to and justifying its proposals. To this argumentation we now turn.

Acknowledgments

The primary recipient of gratitude is Professor José Soltero of DePaul University. José has a natural flair for philosophy and a particular interest in metaphysical matters. His comment on the mathematical consequences of a position concerning the reality of numbers, as discussed in Plato's *Phaedo*, led, by a circuitous route, to the writing of this book. José's patient interest in the structure and applications of the resulting issues, as well as his background in mathematics, was inspirational and instrumental throughout the writing process. On a more human level, his eloquent, dry wit often lightened the relative somberness of our conversations concerning the arguments as they unfolded. In a word, or perhaps three, José Soltero was the sine qua non of this book.

Other individuals were also helpful and must be mentioned. Daniel White combined extensive knowledge of mathematics with equally extensive patience in contributing to the author's understanding of mathematical substance and nuance, as well as providing invaluable assistance with the technological details of book production at the manuscript level. Mary Jeanne Larrabee was also a frequent collaborator in navigating the mysterious eccentricities of computers when employed in the writing of philosophical texts. Unexpected but most appreciated sources of assistance were the efforts of Daniel Gawel to deliver examples of the kind of thinking required for successful completion of problems in group theory. And finally, a note of thanks to Lexington's readers of the manuscript. They combined extensive, informed, and sympathetic critical assessment of the modes of expression for the theoretical blending of mathematics and metaphysics; they also provided a variety of helpful suggestions for enhancing the breadth and scope of the essay's basic argument.

My sincere thanks and appreciation for all that the above individuals contributed to the genesis, development, and execution of this book.

1

Number and Unity in the *Phaedo*

The Platonic Socrates is rarely at a loss for words. However, at a pivotal point in the *Phaedo*, Socrates "paused for some time and was absorbed in thought" (96a).[1] After this period of silent self-scrutiny, Socrates begins to relay his youthful interests in knowing the causes of each thing (including the fact that 1 and 1 somehow "make" 2), an inquiry that he says, in retrospect, elicited considerable inner agitation. Socrates first recounts his critical appraisal of a purely mechanistic theory of causality, which he attributes to Anaxagoras; he then states "the theory of causation which I have worked out myself." This theory is the account based on accepting the existence of Forms and their causal agency regarding the existence of perceptible particulars (100b–103c). Socrates includes in this account a brief analysis of how Forms cause numbers.

A Form is a kind of reality. At various moments in the *Phaedo*, we learn that when Socrates refers to Forms (in the plural since there are many Forms), he intends a reality that is separate, immutable, motionless, and independent from all participating particulars. The discussion outlining how Forms cause numbers, amplified and taken in conjunction with related texts in other dialogues, can serve as a theoretical foundation for raising questions concerning ways to access and assess the mode of existence of Forms in Plato's later metaphysics.

In this book, we consider at the outset a series of implications concerning "the metaphysics of numbers" insofar as the Platonic approach to numbers pertains to metaphysical considerations in general. Alfred North Whitehead's reading of the complex cosmological function of triangles—mathematical entities crucial to the argument of Plato's *Timaeus*—exemplifies speculative possibilities which the this book initially adapts to the range of concerns characterizing the *Phaedo*. This approach to the nature of numbers, when examined and codified in conjunction with a key element in modern abstract algebra, raises a series of important issues regarding metaphysical positions as conveyed in later Platonic dialogues. Furthermore, implications from this investigation in conjunction with certain advances in modern mathematics will yield procedural guidelines (developed in Part II of this book) available for more penetrating analysis of a broad range of metaphysical positions.

The opening modes of analysis in this chapter on the *Phaedo* are therefore rich with possibility and function as a suitable foundation for the speculative adventure to follow.

Numbers in the *Phaedo*—A Play of Paradox

In the prelude to his theory of causality, Socrates reviews several apparent mathematical paradoxes that have stumped him in his quest to master the full range of the concept of causality (96e–97b):

> I who do not even dare to say, when one is added to one, whether the one to which the addition was made has become two, or the one which was added, or the one which was added and the one to which it was added became two by the addition of each to the other. I think it is wonderful that when each of them was separate from the other, each was one and they were not then two, and when they were brought near each other this juxtaposition was the cause of their becoming two. And I cannot yet believe that if one is divided, the division causes it to become two; for this is the opposite of the cause which produced two in the former case; for then two arose because one was brought near and added to another one, and now because one is removed and separated from another. And I no longer believe that I know by this method even how one is generated or, in a word, how anything is generated or is destroyed or exists.

If we write 1 + 1 = 2 and keep the first 1 separate from the second 1, then Socrates cannot understand how a single 1 can become something different, that is, 2, merely because another single 1 is closely conjoined to that first instance of 1. Nor can he understand how, simply because each instance of 1 is placed near the other instance of 1, this proximity somehow "causes" something completely different to exist. Thus, stated more generally, the conjunction (or addition) of distinct instances of one kind of reality does not somehow produce a new reality; rather, the conjunction may be accurately (if abstractly) described as an instance of one kind of thing existing in relation to another instance of the same kind of thing. The proximity (whether established in a spatial or conceptual sense) characterizing the relational existence of the two instances does not inject a factor of determinate increase into this relation that alters or somehow transforms the specific metaphysical character of each instance. The number 1 is, and is only, the number 1.

Socrates aims the same line of criticism against division as the putative cause of 2. An instance of oneness (for example, a single straight line) can be divided into two equal sections. But Socrates wonders how such separation, which he construes as the opposite of the juxtaposition of addition, can produce the same reality—the number 2—when each section of the bisected line is counted and the result is compiled. Generally stated, how can opposite processes cause the same result? But if 1 + 1 does not cause 2, then how does the number 2 come into existence?

The immediate context of inquiry is arithmetic, at presumably an elementary level of operation. However, Socrates' concerns about the origin of the number 2, if amplified, raise intriguing questions with long-range metaphysical repercussions.

Consider the following propositions:

a. $1 + 1 = 2$
b. One plus one are two.
c. One and one causes two.

Propositions (a) and (b) are commonly held to be identical in meaning; they differ only in that (a) is entirely symbolic whereas (b) evokes the symbols in (a) linguistically. This apparent identity makes (c) look odd, at least initially. We might, for example, ask: is "are" in (b) equivalent in all relevant respects to "causes" in (c)? If due attention is paid to the text of the *Phaedo*, the answer is clearly in the negative. "Are" captures the operational aspect of the equals sign; however, if we were to read the "are" as "make," that is, "one and one *make* two," then for Socrates this is seriously misleading since it suggests some sort of active process such that only if we have one and one, and then combine these realities in some undefined way, is it possible to "make" two, that is, the number 2. This is, however, a causal assertion (based on process considerations) Socrates explicitly rejects. Socrates is, in fact, working with a notion of causality developed in a definite theoretical dimension; to appreciate the need for such theorizing (outlined in the following section), we explore the paradoxical consequences Socrates has introduced, albeit in a tightly compressed argument.

The equivalency $1 + 1 = 2$ introduces two additional elements, that is, the number 1 as well as the operation of addition whereby an instance of that number and another, different instance of the same number are juxtaposed with one another. The thrust of Socrates' critique is aimed not just at the individual elements on the left side of the identity statement but also at their grouping, regardless of the operation by which the grouped elements are correlated in order to produce the number 2. Separating the two sides of the equation—$1 + 1$ and 2—introduces the possibility of viewing each side from a broader, more imaginatively but nonetheless critically contoured mode of reference.

Phase A. If unity pertains to each element on the left side of the equation, then unity also pertains to the left-hand side as such, as a single reality encompassing a pair of elements connected with an operator. At this point, the description specifying how this reality is a unity becomes crucial. For if the singleton constituting the left-hand side of the equation represents a kind of totality, a grouping that contains only these two elements, then this unity will fundamentally affect the results of the operation animating that side.

We may now distinguish between (i) the separation of $1 + 1$ from its asserted equality to 2 and (ii) the true cause of the number 2—its participation in the Form duality (according to *Phaedo* 100b, discussed below). This "freeing up" or separation of the left-hand side of the equation from the right-hand side establishes possibilities of interpretive development (which Plato could have anticipated, given that from the appropriate metaphysical perspective, the left-hand side does not "cause" the right-hand side). In sum, the function of *identity* (as indicated above in "a") must be distinguished from the function of *causality* (as expressed above in "c").

We will examine two similar (identity is common to both) but distinguishable ways to generate arithmetically the number 2.

First, consider "1 + 1 = 2 and 3 − 1 = 2." The indicated operations result in a tautology: 2 = 2. As a result, we may write (d): 1 + 1 = 3 − 1. But how does 1 + 1 = 3 − 1 (as such, an identity no less tautologous than 2 = 2) *explain the existence* of the number 2? The equal sign in (d) establishes an equivalence that may be taken as true arithmetically—but only if the number 2 is already assumed to exist! And to account for the existence of the number 2 is precisely the goal of Socrates' inquiry at this juncture in his account of causality.

Now consider, second, the equal sign, the operator in 1 + 1 = 2. Can the = sign, taken by itself, represent causality? If so, then it seems open to propose, for example, that 3 − 1 = 2 or (e) 4 − 2 = 2 are both available as arithmetical routes causing the number 2. (Furthermore, if we allow the entire sequence of integers to exist, there will be unlimited ways to "cause" the number 2.) The second example (e) is particularly intriguing since, although arithmetically correct, the equality uses an instance of 2 (on the left side) as an element in a putatively causal explanation to account for the very existence of the number 2 (on the right side). As a result, this maneuver arguably becomes a *petitio* given that an element in an identity statement becomes a necessary component in order to prove the existence of that very element! Therefore, the alternate arithmetical possibilities for generating 2 seem to point to the fact that the + operator, the plus sign in (a), must be theoretically privileged in establishing, in a uniquely causal way, the existence of a numerical reality, that is, 2, somehow distinguishable from the elements collectively causing that existence, 1 + 1. Notice, as mentioned above, that Socrates' critique aims at not just individual elements on the left side but also at their grouping, that is, regardless of the operation by which the elements are gathered and correlated to produce the number 2.

No other avenue of explanation seems open if the goal is to account causally for the existence of a new number, 2. But, as we shall see in the next section, Socrates follows a different explanatory path.

Phase B. The second phase of the Socratic critique of inadequate approaches to causing the number 2 rests on the same basic argument but with the difference that the element given is a unity—one line—characterized by continuous extension and, as a result, instances a mathematical reality susceptible to being subdivided into two different line segments, each the same length as the other. In sum, "one" line segment + "one" line segment = "two" line segments. But the same criticisms, *mutatis mutandis*, can be aimed at ostensibly causal conclusions generated from this geometrical setting.

The Socratic critique, although played out in a pair of mathematically distinct contexts and thus in a way two-pronged, shares a common strategy; it rejects that the number 2 can be caused by (a) gathering distinct elements in certain ways and then drawing a generative conclusion from this procedure as well as (b) dividing a given singleton magnitude and then inferring that the resulting divisions, when gathered, cause a distinctive and self-subsistent twoness. For Socrates, the process of dividing one thing cannot produce two realities (line segments) which, taken collectively, are distinctively different from the reality exhibited by the original line prior to it being subdivided. In general then, the critique is aimed at gathering or grouping distinct elements (produced in opposite ways—juxtaposing and separating), then drawing identical conclusions about the results of this process. Socrates forcefully denies that either process displays anything like true causal efficacy. Socrates could admit that 1 + 1 = 2 *is equivalent to* 2, thereby safeguarding a basic arithmetical operation, but deny that numerical unity combined with numerical unity *causes* 2.[2]

Numbers in the *Phaedo*—The Paradox Resolved

When Socrates asserts as a matter of principle that he is seeking the cause of 2, it is crucial, given the dialogue's overall concerns, to keep in mind that the appeal is to "cause" understood in a technical, theoretical sense. At *Phaedo* 99b, Socrates draws a fundamental distinction: "in reality a cause is one thing, and the thing without which the cause could never be a cause is quite another thing." Socrates cites his own circumstances to illustrate the distinction. The cause of Socrates sitting in jail awaiting execution is not his body and the fact that it remains motionless awaiting the arrival of the warder bearing the poison—the cause is that Socrates has moved his body into this position (waiting patiently for his death) because he was convinced that it was "for the best" that he acts this way.

The Platonic Forms now appear, or more precisely, reappear. At this juncture, Socrates reminds his audience that what he is about to say about the Forms is nothing new, given that they appeared earlier in the day's discussion (74a, 76e) as well as in prior dialogues. Thus, if they will "agree that these [the Forms] exist, I believe I shall explain cause to you and shall prove that the soul is immortal" (100b). Pursuing a generalized account of the nature of causality is therefore a prerequisite for what has been the primary philosophical issue of the day—demonstrating by argument the immortality of the soul, an obvious matter to consider given that Socrates' execution will take place at sunset (with lingering questions and concerns as to the effects of death on his soul). After briefly recapitulating his critique of the earlier attempt to generate the number two, Socrates forthrightly states what, for him, is the correct explanation:

> Well, then, if one is added to one or if one is divided, you would avoid saying that the addition or the division is the cause of two? You would exclaim loudly that you know no other way by which anything can come into existence than by participating in the proper essence of each thing in which it participates, and therefore you accept no other cause of the existence of two than participation in duality, and things which are to be two must participate in duality, and whatever is to be one must participate in unity, and you would pay no attention to the divisions and additions and other such subtleties. (101c–d)

This is the classic passage announcing that numbers, such as 2, depend for their existence on a causal connection to something more fundamental, that is, the Form two or duality. It should be pointed out, however, that a later discussion in the *Phaedo*, the final proof of the soul's immortality, adds (in context as a step in this proof) an important feature to what the Form of twoness confers on instances of the number 2 (and, *pari passu*, all numbers): "when the form of three takes possession of any group of objects, it compels them to be odd as well as three" (104d). We may infer then that a group of three things is odd because the number 3 is odd, just as two things are even because the number 2 is even. This addition is surely of metaphysical significance, since Forms for numbers not only confer the reality of the numbers as such, in and of themselves, but also *important mathematical relations between a given number and other numbers of a different order*. (This dimension of the causal activity of numerical Forms will be discussed later in the book.)

Prior to stating the more adequate account of what causes the number 2, Socrates had registered a terse but broad-ranging criticism of extant theoretical approaches to causality (99c)—"in truth they give no thought to the good, which must embrace and hold together all things."[3] What precisely happens, and presumably does not happen, by virtue of the good as it "holds together" all things is not elaborated in the *Phaedo*. And it is because Socrates was unable to learn about the nature of the good, either by himself or through instruction from others, that he proposes to follow a "second sailing" in this explanation of causality (99d), an apparently derivative account that controls the remainder of the day's conversation (including the need to relate the eschatological myth after the various arguments for the soul's immortality have concluded). However, it is possible to extrapolate certain implications from the account of numerical causality Socrates has just provided. And as we shall now see, the senses in which the good, although inchoate in the *Phaedo*, can function at a foundational level in "holding things together" are crucial with regard to broader connections with mathematical elements.

Mathematics, Forms, the Good—An Introductory Survey

The development of the notion of the good in the *Republic*—as well as a speculative development of this notion—will become a dominant theme in Chapters 7 and 8. Some attention to the place of the good as initially stated in the *Phaedo* is therefore justified at this juncture.

i. Socrates and the attendant company have assumed the existence of the Forms. But the Form unity fundamentally differs from the Form duality—each Form is one Form, but what the oneness of the Form unity confers differs from what the oneness of the Form duality confers. The difference is, so to speak, in the internal power of bestowal as far as the effects of the two Forms on particulars are concerned. The Form unity causes the number 1 to exist as that number. This difference in causal effect does not, for example, justify claiming that the Form duality is twice as powerful as the Form unity (since, putatively, the former affects double the number of related particulars compared to the latter). We will assume then that no one Form, as a unity, is any more or less causally powerful than any other Form. We can, however, assert that the number 1 differs from the number 2 not only in ways open to elementary arithmetical computation; rather, more fundamentally, 1 is a different kind of reality than 2 since the two numbers derive their reality from Forms which themselves differ from one another both as such, with respect to their inherent, substantial constitution, as well as in terms of the type of reality each Form confers on a particular.

Each number, 1 and 2, exists in itself (because of a relation each number has to a Form), and it may be assumed at this juncture that the distinctive identities of each number will remain as they are, as well as the numerical sequence 1, 2 maintaining this sequence and stated in this order—all because the good operates at this basic level of causal and arithmetical reality.

ii. All groups of two things, dyads, are held together as dyads by the Form of duality. Thus, if duality causes the number 2, then it may be inferred that duality causes any grouping that has two elements. If so, then duality (as a Form) confers (a) the character of twoness on a single group of distinct particulars as well as (b) the fact that the particulars exhibit a unity, that is, they constitute a single instance of twoness.

iii. The duality of each set of two elements may be identical with respect to the number 2 by itself (as the number 2) and prior to its use in referring to non-numerical entities (a group of two things). But that any two distinct particulars can be "gathered" together to instantiate the predication of the number 2 represents a distinct function of the number 2 as an instrument of duality when, as that number, it initiates and affirms control over the realm of particularity with respect to this grouping function.

iv. By parallel reasoning, any Form that causes the character of having a number also confers and maintains (or holds together) the same kind of grouped unity. But we may now point to the distinct number of elements as the relevant constitutive feature within the confines of a single, given group (i.e., a group comprised of any number of elements after the number 1). Furthermore, this feature underlies the possibility that the elements of a group can be arranged with respect to one another in a wide variety of orders, all of them deployed, or held together, within the circumscribing limits of the group.

v. As a matter of metaphysical fact, the good holds together an extensive variety of dyads; by contrast, however, it is logically possible that such minimal grouping may never have been metaphysically possible. In the latter case, reality would be a continually unpaired mingling of individual entities, none of which would have anything more than the briefest—and random—interaction with any other entity. In fact, all manners of groupings characterize the immense multifariousness of what we are pleased to refer to as the natural order (indeed, the phrase "natural order" itself reflects the complex and diversified intersection of these groupings). But why must this fundamental level of paired "sociability" be the case? For Socrates, the ultimate cause is the good in its capacity to generate a wide variety of ordered connections and relations between Forms (those Forms conferring numerousness) and particulars.

vi. Dyads constituted by distinctive particulars of the same type could also be constituted by particulars marked by differences—for example, a pair of socks vs. a pair of athletes (an archer and a wrestler), or two friends who have little in common but who still like each other, and so forth. The formative agency differs because the kinds of reality formed differ with respect to each other—thus two socks (whether matching or non-matching) as opposed to two athletes exemplify two different kinds of twoness. In addition, the dyadic quality of pairs of things will vary in terms of what must come into play in order to gather different particulars, in some cases very different particulars—for instance, a pair of shoes in contrast with a pair of lovers—so that such particulars instantiate individual instances of twoness constituted as such by a measure of duality "in" and thoroughly permeating the two entities as a pair.

vii. Does twoness or duality in some way incorporate two distinct entities (even if as a duality they are taken as one uniform set)? This and similar questions revolving around the notions of unity and plurality become important, if not crucial (especially for Aristotle, as we shall see in Chapter 5).

Another, related metaphysical aside: it will be necessary at some point to pursue additional abstract aspects of pairs constituted as dyads. Thus, the component of time is relevant, "durational duality" as it might be called. After all, couples break up, and one member of a pair of socks can, on occasion, seemingly go off and hide. Since the factor of twoness as such does not change, particular instances of duality that lose this factor will have to be analyzed in order to determine the connection between particulars existing separately and existing as a pair, and the fact that the latter state is subject to change.

viii. The above considerations illustrate how the good holds together elements with respect to the Form duality and its explicitly stated function regarding particular instances of twoness. But the good also holds the integers together in a given sequence—thus one is followed by two, two is followed by three (note all the references to the number three at *Phaedo* 104a-e). By virtue of this aspect of the good, each integer exists both by itself, as a distinctive number, and the integers are also held together collectively as members of an ordered sequence which, as such, is presumably fixed and not subject to variation. This double mode of existence is conferred on each number by virtue of sharing in a Form, a type of entity that always exists—immutable, invisible, and one.[4]

The Socratic interest in a theory of causality proposes an account that does considerable work, especially with respect to the existence of numbers—individually, as numerically formative agents, and as members of a fixed series. However, at this juncture, we might register the critical concern, surely to be voiced by some mathematicians, that we have been introducing substantial content into a purely formal representation of the basic elements (as well as basic operations) in arithmetic as well as a kind of elementary number theory. Socrates might reply that this is precisely one of the metaphysical goals of the inquiry, that is, to inquire into the meaning and cogency of an appeal to "formal" when predicated of individual numbers as well as the results of operations, each taken as a kind of formal unity when they employ those individual numbers. At present, this programmatic merger of mathematics and metaphysics may nonetheless appear to some observers as little more than imaginative proclivities threatening to run theoretically amok. In that regard, we may note that numbers function as a framework of immediate stability, especially from the standpoint of their existential practicality in the quotidian activities of human affairs. In this respect, then, numbers and their apparently inherent stability may serve as paradigms for many other types of reality that exhibit at least a measure of such stability, if not the apparently full-fledged immutability of numbers themselves.

An attitude of caution at this point and prior to proceeding is nonetheless prudent. Once inquiry elicits at a deep level the metaphysical foundations for mathematical realities, the possibility of extensive ramifications in both areas becomes evident. The argument in this book will be that the apposition of mathematical operations with metaphysical elements produces useful perspectives for appreciating relevant aspects of those elements. This interpretive possibility suggests a way to take an intriguing and, perhaps, cautionary detail appearing in the statement of Socrates' theoretical ventures. Socrates asserts that anyone involved in a conversation concerning the generation of numbers would "exclaim loudly" (μέγα ἂν βοώης 101c2) the general principle that "you know no other way by which anything can come into existence than by participating in the proper essence of each thing in which it participates."

This claim represents a classic statement of the Platonic position on participation, that is, that any particular comes to exist by participating in its originating Form. However, we may be allowed to wonder why Socrates insists that the principle be exclaimed loudly by its advocate. After all, the adequacy of a fundamental principle of metaphysics hardly depends on how loudly that principle is announced. One interpretation of this textual detail is that Socrates wants the principle to be taken seriously as a more adequate counterpart to the deficient account based on an unconsidered reliance on causality—simply put, when philosophizing in conversation, the louder the better with respect to articulations of axiomatic metaphysics. However, a more penetrating (but also more speculative) possibility is that Plato is alerting the reader, albeit subtly, to the fact that this principle may require additional finely grained scrutiny precisely because it is so crucial in the present context, that is, to the approach to the generation of numbers Socrates is advancing. Such analytic care would be all the more pertinent given the varied and complex issues raised when inquiry goes into depth concerning metaphysical considerations pertaining to the generation of numbers.

The sections of the *Phaedo* analyzed in this chapter mark a fertile intersection—elementary mathematical notions and operations aligned structurally with a complex metaphysical foundation. The metaphysical vision generated by this intersection is impressive, even if its full scope stands inchoate at this point. However, side by side with this positive expectation are diverse questions and problems that must be raised and which, upon reflection, temper any initial attraction the Platonic approach to numbers may seem to engender. Aristotle articulates an array of these problems, and some of the most penetrating and pressing of them will be reviewed (Chapter 5), analyzed in related metaphysical contexts (Chapter 6), and then integrated in positive ways with the programmatic objectives presented in the book's final two chapters. But now, while still in the early going, we will introduce another dimension of the revelatory mathematics/metaphysics nexus by examining a perspective on Plato asserted by a contemporary thinker who was both a mathematician and a metaphysician.

Notes

1 Harold N. Fowler's translation in the Loeb *Phaedo* (Cambridge, MA: Harvard University Press, 1977). Translators vary in rendering the temporality denoted by συχνὸν at *Phaedo* 95e7. Hugh Tredennick also has "some time in reflection"; however, R. Hackforth and G. M. A. Grube both have "paused for a long time." The relevant point, given the typically easy flow of Socratic conversation, is that Socrates ceased talking altogether and had to think in silence about the issue at hand, regardless of whether this enforced reflective exercise required some

time or a long time. Plato alerts us to the fact that what the temporarily silent Socrates will say to his associates is crucial, complex, and drawn from the depths of his soul. Translations from the *Phaedo* are Fowler's unless otherwise indicated; translations from other dialogues will be identified when they occur. On occasion, a translation has been modified and will be so indicated. Citations for the translations mentioned above: Tredennick is in *The Collected Dialogues of Plato*, ed. Edith Hamilton and Huntington Cairns (New York: Pantheon Books, 1961); R. Hackforth, *Plato's Phaedo* (Cambridge: Cambridge University Press, 1972); G. M. A. Grube, *Five Dialogues,* 2nd ed. (Indianapolis: Hackett, 2002).

2 For a related discussion of Socrates' implied criticisms for determining the origin of numbers in the *Phaedo*, see Reginald E. Allen, *Plato's Parmenides,* Revised ed. (New Haven: Yale University Press, 1997), 96–8.

3 Fowler takes the reference to "the good" (τὸ ἀγαθὸν–99c5) in this sentence in a strong substantive sense. A review of other translations is instructive. In Benjamin Jowett's rendition: "of the obligatory and containing power of the good they think nothing; and yet this is the principle which I would fain learn if any one would teach me." Jowett preserves, correctly in our view, the substantial character of the good. R. Hackforth: "for they don't believe in any good, binding force which literally binds things together and holds them fast." For Hackforth, the good becomes dissolved into a descriptor modifying "force," thereby weakening the metaphysical substantiality of the good. Hugh Tredennick: "they do not think that anything is really bound and held together by goodness or moral obligation." Tredennick sees "the good" as equivalent to "goodness or moral obligation." The appeal to "moral obligation" captures the appropriate normative dimension of the passage but tends to downplay the sense that by means of the good, everything is "bound and held together." Finally, G. M. A. Grube: "and they do not believe that the truly good and "binding" binds and holds them together." This version seems to separate "truly good" from what is "binding," thereby diminishing the substantive force of "the good" (unless Grube is using "and" in "truly good and binding" as the "explicative *kai*," that is, "truly good *and therefore* binding"). The interpretive issue concerns the extent to which "the good" should be understood as a fundamental element in the metaphysical structure Socrates is developing as a necessary condition to account for causality. In our view, Fowler's version as it stands most accurately captures the intended sense—see also in this regard Note 4.

4 D. Roochnik outlines how the act of counting helps turn the soul around and "shift one's sights" toward the good. See David Roochnik, "Counting on Number:

Plato on the Goodness of *Arithmos*," *American Journal of Philology* 115 (1994): 543–63. In Heath's phrasing, mathematics goes hand in hand with philosophy and "the value of the two sciences consists in the fact that they draw the soul toward truth and create the philosophic attitude of mind, lifting on high the things that our ordinary habit would keep down." See Sir Thomas Heath, *A History of Greek Mathematics*, Vol. 1. (New York: Dover, 1981), 284. For a systematic treatment of the good in the *Phaedo*, see David A. White, *Myth and Metaphysics in Plato's Phaedo* (Selinsgrove: Susquehanna University Press, 1989), 153–75, 185–204.

2

Whitehead and Plato

The Legacy of Triangles and Number

As a metaphysician, Alfred North Whitehead (1861–1947) was influenced by Plato in a variety of ways. Most prominent perhaps was borrowing what Whitehead felt were relevant features of Platonic Forms and adapting them for his own purposes so that the results—especially as found in *Process and Reality* (1929)—cohered with scientific advances Whitehead experienced in his world. An obvious case in point: the diverse complexity of Whitehead's notion of "eternal objects" invites critical comparison with canonical Forms, especially as depicted in Plato's middle period dialogues. However, the extent to which doctrinal Whitehead is or is not a Platonist is primarily a historical question and will concern us here only in ways that pertain materially to the basic goals of our inquiry. Chapter 2 reviews a set of Whitehead's reflections on the significance of positions involving mathematics as stated directly in Plato's dialogues. This account will contribute a modern counterpart to Plato as well as foundational support for the more speculative adventures pursued later in this book.

Whitehead on Plato—Mathematics and the Structure of "Natural Things"

Whitehead's gnomic summary of Western philosophy, that it consists of a series of footnotes to Plato, is well known. Perhaps not as prominent is his assessment of Plato's metaphysically incisive and perceptive vision of determining the underlying structure of natural phenomena by applying features proper to basic geometrical entities—triangles, as well as the function of integral numbers in musical harmony. Here is Whitehead on Plato as a kind of mathematical and metaphysical "prophet":

> Newton could have accepted a molecular theory as easily as Plato, but there is this difference between them: Newton would have been surprised at the modern quantum theory and at the dissolution of quanta into vibrations; Plato would have expected it. While we note the many things said by Plato in the *Timaeus* which are now foolishness, we must also give him credit for that aspect of his teaching in which he was two thousand years ahead of his time. Plato accounted for the sharp-cut differences between kinds of natural things, by assuming an approximation of the molecules of the fundamental kinds respectively to the mathematical forms of the regular solids. He also assumed that certain qualitative contrasts in occurrences, such as that between musical notes, depended on the participation of these occurrences in some of the simpler ratios between integral numbers. He thus obtained a reason why there should be an approximation to sharp-cut relations of harmony standing out amid dissonance.[1]

A note on the interpretive context underlying Whitehead's tradition-hopping comparison and subsequent inference will help introduce a strategically parallel approach (developed in this book) to the Platonic treatment of numbers. Just as Plato was ahead of his time with respect to geometry and the import of regular solids formed by triangles when their structural features are juxtaposed with modern chemistry and physics, we will show that he was also ahead of his time if the account of the causality of numbers is placed in apposition with fundamental elements in Plato's later metaphysics (outlined in Chapter 3) and then juxtaposed with a seminal configuration in modern abstract algebra (Chapter 4).

What led Whitehead to cast the Plato of the *Timaeus* in the role of metaphysical prophet? For Whitehead, Plato's *Timaeus* and the *Scholium* of Isaac Newton represent "the two statements of cosmological theory which have had the chief influence on Western thought" (PR, 93). In fact, comparing the two documents reveals that what the *Timaeus* "lacks in superficial detail, it makes up for by its philosophic depth"; indeed, if "read as an allegory, it conveys profound truth" (PR, 93). The following sketch of key sections of the

Timaeus draws from the translation and commentary of F. M. Cornford, a standard secondary work on this dialogue.[2]

If fire, air, water, and earth are accepted as natural "givens," then the god shaping the universe—a divinity whose existence Plato assumes as a cosmological necessity—gave each of these elements "distinct configuration by means of shapes and numbers. That the god framed them with the greatest possible perfection, which they had not before, must be taken, above all, as a principle we constantly assert" (198/53b). This framing gave each element "the shape of one of the regular geometrical solids, so that there are just four shapes, chosen because they are the 'best'" (210/ 53c). The four shapes are in turn constructed from triangles, and Cornford identifies the most basic types: "This right-angled isosceles triangle and the half-equilateral (which is one type out of the infinite variety of right-angled scalene triangles) are thus taken as the two irreducible 'elements' for the construction of all the four solids" (211/53c-d). In addition, the triangles differ in size, and when this factor is taken into account, the result is "an endless diversity" of natural things and phenomena "which must be studied by one who is to put forward a probable account of Nature" (230/57d). Finally, the expansion of *Timaeus* 53d6-7 neatly presages a principle of gathering which incorporates mathematical and metaphysical consequences of fundamental importance. Cornford writes: "Philosophic mathematicians...will know that a very long chain of propositions reaching back to the ultimate premisses of the science is needed before you can solve problems so difficult as the construction of the regular solids." (213/53d). In subsequent chapters, we introduce elements pertaining to these "ultimate premises" of mathematics in ways that extend the import of Whitehead's approach to Platonic positions regarding the basic building blocks of the cosmos.

Here we see a mathematician/philosopher looking at a philosopher's use of mathematical elements and reading this usage as a creative anticipation of a theory in modern science not broached explicitly until two millennia after the philosopher had applied mathematical elements toward philosophical ends. Whitehead juxtaposed the text of the *Timaeus* with modern scientific developments and then proposed an imagined frame of reference that sanctions reading the Platonic text so that it is possible to discern elements that underlie both modern and Platonic positions. Whitehead read the Platonic use of triangles—and all the mathematical factors associated with this basic geometrical element—as an allegorical anticipation of molecules. Instead of advancing an immediately perceived acceptance of the "sharp-cut differences between kinds of natural things," Whitehead emphasizes the fact that Plato discerned natural things, at their most fundamental, as geometrical realities capable of numerical measurement. As such, these elements were, for Whitehead, "approximations" of the modern use of molecules characterized by a common structure. In the Platonic scheme, these triangles, or basic elements, were capable of unlimited interaction in such a way as to produce the "endless diversity" of Nature. The introduction of triangles signaled Plato's recognition that natural things exhibit abstract, general measures of order to the extent that the underlying structure of that order is expressible through mathematical configurations.

Whitehead's reading went from mathematics to physics, but within a thoroughly metaphysical context of investigation; this approach allows Whitehead to contend that Plato anticipated pivotal features of reality theorized in subsequent areas of scientific inquiry. We will show that the same kind of imaginative prefiguring can apply to the Platonic treatment of the generation of numbers, that is, that this treatment contains elements that anticipate advances made in modern mathematics, in particular a theoretical, duly axiomatized dimension of an important branch of modern algebra. If therefore the context of inquiry shifts from triangles to numbers, from geometry to arithmetic, we will see that (a) Whitehead's juxtaposition of mathematical elements in the *Timaeus* with apposite positions in modern physics and chemistry establishes an illustrative basis for (b) correlative interpretive comparisons involving arithmetical positions adumbrated in the *Phaedo* when these positions are analyzed in conjunction with a series of passages on methodology in later dialogues, and then juxtaposed with certain fundamental themes in abstract algebra.[3] Chapters 3 and 4 develop these themes.

Whitehead's example of Platonic mathematical theorizing and its incisive consequences for subsequent thought concerning the metaphysical structure of natural things is an important historical feature of metaphysical concepts and positions laid out in *Process and Reality*. However, Whitehead's interpretive contributions in natural contexts are complemented by two areas of additional reflections relevant to this kind of speculation. First, consider the *Timaeus* briefly and indirectly evoking the good in a mathematical context, that is, the appeal to "the best" as far as the structure of certain kinds of triangles is concerned. A notion of the good, derived from the Platonic approach presented in the *Republic* (but considerably amended and redirected), becomes increasingly prominent to the general argument of this book. And it is worth noting by way of preamble that Whitehead's late essay, "Mathematics and the Good," will be integrated into the treatment of the good in Chapter 7. Second, Whitehead contributed another, quite distinct perspective on Plato and mathematics in his *Modes of Thought.* A brief summary and discussion of this perspective will extend Whitehead's significance as a foundational figure regarding the primary goals of this book.

Whitehead on Plato—Numbers and Process

Whitehead's most extensive speculative work, *Process and Reality,* appeared in 1929. A decade later, Whitehead published *Modes of Thought*, a compilation of lectures delivered on various occasions after *Process and Reality* and intended, as he says in the Preface (vii), to condense features of his thought "incompletely presented in my published works."[4] The approach to Plato taken in *Modes of Thought* emphasizes numbers rather than geometrical figures and differs considerably in interpretive direction from the position outlined earlier and discussed in the previous section.

The Whitehead of *Process and Reality* is generally positive regarding Plato's influence on subsequent philosophical thought; however, *Modes of Thought* presents a much less sanguine approach to Platonic principles. In the later work, Whitehead emphasizes what he sees as "the erroneous notions of process devoid of individuation, and of individuation devoid of process" (MT, 96); furthermore, he insists that the western tradition has followed the lead of Plato, with the position stated in the *Phaedo* (and discussed in Chapter 1) a primary source of this error. Here is the relevant section quoted at length (MT, 96–7), a passage in Chapter 5, "Forms of Process":

> The notion of number, as elaborated in arithmetic, has been traditionally treated with this bias toward such an erroneous separation. Each individual thing is devoid of numerosity, whereas, a static group is characterized by number. In this way *process* [my italics] seems to be absent in our treatment of arithmetic. Thus mathematics has been conceived as the test case, which is the citadel for a false metaphysics.
>
> When Plato thought of mathematics he conceived of a changeless world of form, and contrasted it with the mere imitation in the world of transition. Yet when Plato thought of the realities of action, he swayed to the opposite point of view. He called for life and motion to rescue forms from a meaningless void.
>
> In these lectures [*Modes of Thought*] Plato's second doctrine, of life and motion, has been adopted. The mathematical modes of fusion, such as "addition," "multiplication," "serial form," and so on, have been construed as forms of process. The very notion of "multiplicity" itself has been construed as abstraction from the form of process whereby data acquire a unity of issue into a novel datum.

Much of this passage is worth close and reflective examination. For now, we note that Whitehead distinguishes sharply between a "changeless world of form" and "the realities of action"; furthermore, he insists that when Plato had the latter under philosophical scrutiny, he "called for life and motion to rescue forms from a meaningless void." Mathematics, and numbers incorrectly conceived—and therefore standing as "the citadel for a false metaphysics"—appear clearly and unavoidably present given Socrates' claims in the *Phaedo* regarding the number 2 (and all numbers) being caused by Forms construed in their canonic state, that is, separate, immutable, motionless, and independent from all participating particulars.

Before we explore implications from this position, we may be allowed a brief rejoinder. Examination of the *Phaedo* shows that Socrates asserts the following as a matter of principle: it is precisely by virtue of access to the Forms, insofar as their presence resides in the soul, that we should lead our lives—presumably in as many lived contexts as possible—as if the body is a "prison" (62b4). At this point in the dialogue, Socrates is not at all disposed "to rescue forms from a meaningless void"; indeed, the ongoing philosophical quest in the *Phaedo* is to emphasize that the pervasive quality of the Forms throughout the complex fabric of our lives is precisely what makes life truly meaningful for the soul. Only if the soul can win the appropriate attitude toward the body—that it is a veritable prison with regard to accessing true reality—can the person, as an amalgam of soul and body, live in a properly philosophical manner. Although Plato modified this "puritanical" strain in his ethical positions in later dialogues, his texts never retreated from the importance of Forms as far as the actual course of life, to use Whitehead's term, was concerned.

The point is not to interject a hit-and-run criticism of Whitehead's adequacy as a Plato commentator; rather, it is to suggest that given the goals of this chapter, it will be far more productive if we follow Whitehead's own proposal after his "footnotes to Plato" dictum with respect to the epochal influence of Plato on subsequent Western philosophy. Whitehead immediately adds at this point that he is not referring to "the systematic scheme of thought which scholars have doubtfully extracted from his writings," but rather "to the wealth of general ideas scattered through them" (PR, 39). The present strategy, bluntly stated, is to do with Whitehead what Whitehead says so many thinkers have done, with fruitful results, employing elements of Platonic thought.

On the Limits of Speculation: Identities and Diversity

When juxtaposed with one another, Whitehead's positions on Plato display a stark interpretive contrast. The reading in *Process and Reality* of the explanatory power resident in mathematical entities (such as triangles) in the *Timaeus* is positive; the approach in *Modes of Thought* to static formal realities such as the Forms underlying the existence of numbers in the *Phaedo* is negative. Indeed, the two interpretive perspectives on Platonically rendered mathematical entities may appear initially antithetical to one another in such dramatic logical fashion that any attempt to forge a theoretical conclusion based, in part, on Whitehead's support would seem obviated from the very outset. Is there some way to combine the two divergent approaches to Platonic views on mathematical realities so that they jointly indicate, or at least hint at, metaphysically relevant possibilities? The answer is in the affirmative.

Not long after confronting the striking image of a "citadel for a false metaphysics" established under Platonic supervision and control, we discern possibilities for notable, useful development. Consider the following passage (MT, 98):

> Thus the differences arising from diversities are not absolute. Analogies survive amid diversity. The procedure of rationalism is the discussion of analogy. The limitation of rationalism is the inescapable diversity. The development of civilized thought can be described as the discovery of identities amid diversity. For example, the discovery of identities of number as between a group of days and a group of fishes.

Whitehead appeals to "the discovery of identities amid diversity" as a product of developed "civilized thought." Although "civilized thought" may not sit well with contemporary inclusive cultural sensibilities, the neat and more abstract evocation of discovering "identities amid diversity" raises the question of whether the identities so discovered (regardless of underlying cultural circumstances) will occupy residence in the Platonic metaphysical scheme, or whether they must be located in another theoretical framework.

As we have seen, the imposed dichotomy between life and the Forms emerging from the Whitehead of *Modes of Thought* sharply separates these two modes of reality from one another. This stark difference is not only inaccurate (if not patently false) with respect to the text of the *Phaedo,* but it is also at least potentially detrimental to the full philosophical import of each domain, the Forms and life, in relation to each other—a limiting consequence on theoretical possibilities which could well be at odds with Whitehead's own metaphysically foundational aspirations. As Whitehead himself says in this context (MT, 99), "The notion of potentiality is fundamental for the understanding of existence, as soon as the notion of process is admitted." Even at first glance, this broadly expansive assertion linking *potentiality* with *process* seems open to a mode of interpretation manifesting a certain level of imaginative courage.

With regard to the "notion of process," it should be noted that "identities amid diversity" taken as a sort of metaphysical slogan sanctions a variety of possibilities, especially with respect to the referential content of "identities." Thus, a Platonic Form is an identity in a very strong sense, since its existence remains what it is throughout any and all modes of possible predication (and, for Whitehead, its blockage of process testifies to its overly rigid and thus "unreal" status as a fundamental element in the description of reality). By contrast, an entity could enjoy an identity for a certain period of time and under a certain set of circumstances, but then become altered (thus involving itself with various sorts of process). An anticipatory note: We will have occasion to return to this broad-based locus of metaphysical speculation at various junctures in Part II of this book. We shall follow Whitehead's main thrust in this speculative arena and introduce a variety of speculative potentialities and possibilities emerging from the interplay of Forms and process. These approaches will be drawn from additional and, when feasible, redefined as well as redeployed Platonic sources.

For now, we are invited, or perhaps "challenged" is more appropriate, to rethink the reality of process as it characterizes (or, with Whitehead, should characterize) the realm of mathematics. Consider, as a point of departure, the existence of numbers. How does process pertain to the mode of existence displayed by a number such as 2? Whitehead illustrates the discovery of "identities of number" by identifying groups of different things—fish and days: for example, two fish and two days. The diversity marking these two types of reality is manifest; however, permeating this diversity is an identity—the number two. For the Platonist, this identity has been established by a reality—2 (or duality)—which (a) can be separated from the things that the reality has enumerated and (b) exists by itself, in its own right, displaying a mode of existence inherently separable from particular things.

For Whitehead, such a venture into metaphysical theory, even if generated from an apparently formal mathematical base of operation, violates the necessary interplay of process and stability (taking stability as a threshold characteristic of identity). The force of the distinction just introduced between (a) and (b), and the consequent elevation of the resulting realities to a realm presumably apart from any dimension of process, would fall within theoretical confines artificially (and classically) fashioned, according to Whitehead's evocative image, as "the citadel of false metaphysics." But the question raised above, concerning the sense or senses in which process can apply to numbers themselves, continues to be a pressing concern.

Process and Mathematics: A Schematic Preamble

It is worth examining the extent to which Platonic positions exemplifying a high degree of formality necessarily reduce to measures of metaphysical falsity in this context, especially from the perspective of possibilities derivable from mathematical factors and discoveries. For as we shall see, the stone-like impenetrability conveyed by Whitehead's conjectured "citadel" does in fact yield certain openings of accessible development.

In *Modes of Thought*, Whitehead construes addition, one of the mathematical "modes of fusion," as an instance of the "forms of process." The term "fusion" is multifarious in the natural order as well as in human affairs, for example, currently in use to describe the formation of basic elements within the inner workings of certain types of stars as well as to identify a kind of popular cuisine. Whitehead's initial setting for the subject matter of this sequel to *Process and Reality* was a public lecture forum: thus, allowance for flexible, malleable terminology—for example, "fusion"—dealing with inherently abstract, complex subjects may be appropriate discourse for such a voiced narrative and literary setting. However, the metaphysician appreciates precision and referential certainty, assuming that these characteristics are in fact available. As an initial step in the quest for such terminological substantiation, it is surely evident that fusion, as used here, is intended to evoke process elements, ultimately introducing dimensions of possibility in mathematics at the level of basic arithmetic and, by extension doubtless in even more sophisticated branches of this discipline. How then to seek out verbalized accesses to these dimensions?

Generally speaking, fusion denotes a process by which something new is generated from something that had already existed. Recall that Socrates denies that addition, a type of fusion for Whitehead, can produce something new—thus the fusion of 1 and 1 cannot cause 2; yet for Whitehead, addition is necessarily a type of fusion insofar as it instantiates a form of process, an essential component in this kind of analysis. The contrast with respect to ultimate metaphysical function could not be more stark. The primary context of inquiry can be localized in a Whiteheadean framework if it is possible either (a) to translate some, or any, of the desired senses of fusion into traditional metaphysical elements and settings or (b) if necessary, to coin new language (duly defined and clarified as to usage) to evoke in process terminology what we may assume for now (and with Whitehead) is a requisite metaphysical desideratum.

Although this book's argument begins with an examination of the metaphysics of numbers, its ultimate goal is significantly more comprehensive in scope—to appreciate and articulate the significance of mathematical considerations in the practice of metaphysics. Given the dramatic differences between the two philosophers considered so far, we will proceed by approaching the phenomenon of process with explicitly intended measures of sympathy but in a broadly circumspect manner. To this end, we conclude the first two chapters—the confrontation at a high level of generality between Plato and Whitehead—with a series of types of process (illustrated with positions already discussed in this book) that can serve as heuristic guidelines for subsequent investigation. The impetus for such an interpretive schematic is Whitehead's principled insistence on process as a pivotal (and, if he is correct, often underrepresented) factor in metaphysical accounts, even and especially the rarified dimensions of reality housing the elements and propositions of mathematics. This slate of settings (we offer four possibilities) for the determination and inculcation of process factors arising within the intersection of mathematics and metaphysics becomes increasingly more relevant as we proceed.

Relational

Assume that generally speaking, any instance of fusion is, initially, a relation (a) originating from elements distinct from one another in some way (or ways) and then (b) marking the intersection of these elements through a type of merger or combination that produces a new result, presumably with an identity not reducible to the identities of either (or any) of the original elements. In general, *relational* process refers to modes of interaction between and among particular elements.

This approach to process rests on an assumption, that is, that elements or particulars which at some point participate in this type of process possess an established identity of their own prior to the onset of such modes of interaction. The challenge at this stage of the sought-for phenomenon of process will be to invoke language capturing process when it is "on the move," so to speak, between two states—the condition of the elements prior to the onset of fusion and the result of the fusion once that process has been completed.

However, consider this possibility: what if process is *always* present and, idiomatically stated, always on the move? If so, then without additional qualification it would seem that any asserted identity is and must remain only temporary. Thus, the specter of unending, doubtless turbulent Heraclitean flux hovers over all such reflection. The notion of *limit*, a limit of any sort, should be introduced only with care and precision within such a domineering flux. *This* fish is not *that* fish, but the determination of such an apparently basic difference, the limits of each fish, each entity with respect to everything else, must be open to some form of articulation in order to establish a measure of stability, a temporary pause in the onrush of seemingly unchecked process. The next dimension of process to be considered in this heuristic preamble becomes crucial in this regard.

Substantial

Recall Whitehead's claim quoted above: "Each individual thing is devoid of numerosity; whereas, a static group is characterized by number." We will sketch heuristic concerns applicable to determining senses of *substantial* process proper to an "individual thing" as well as a "static group," a quite distinct type of reality.

Assume that Whitehead is correct and that addition, if construed in terms of metaphysical discourse, is not a purely formal or axiomatic exercise but instances an especially fundamental kind of relation, that is, a fusion of sorts. The following question then seems pertinent, if not unavoidable: what happens, if anything, to, for example, the number 1 if it is "fused" with another instance of the same number in order to become the number 2? This question generates a series of related and equally basic (if not more so) questions. We outline several points of investigative departure at this juncture in the following schematic preamble—first, with respect to individual things (a-c) followed by, second, static groups (d):

Individual Things

a. The question as posed assumes that 1 is a number and, as such, that it can identify "each individual thing" as an "each" and as an "individual." However, the status of 1 as a number for the Greeks of antiquity—our initial locus of inquiry—is disputed. Furthermore, as noted above, Whitehead appears to contribute to this debate when he asserts that "[e]ach individual thing" is separated from or devoid of "numerosity." If we take "numerosity" in what seems to be the intended context, then it appears as if an individual thing lacks the capacity to be referred to by an existing number; if so, then it also seems to follow that 1 is not a number—since, as just noted, any single individual thing could presumably be identified as *one* thing (regardless of any type considerations, whether the type is obvious or indeterminate).

Consider the apparently tight connection between 1 as a number in mathematics and the notion of unity in metaphysics. Recall in this regard that Socrates straightforwardly asserts that "whatever is to be one must participate in unity" (*Phaedo* 101c—cited in Chapter 1). The denial of 1 as a number would seem to imply then that an entity could exist as a singleton, *one* entity (e.g., one fish), without simultaneously also having the capacity to be construed arithmetically as displaying this numerical mode of existence as an element integral to its reality—surely an eccentric confluence of senses of unity. Therefore, even in what appears to be a narrowly defined context of inquiry, the status of unity appears to be complex, especially regarding apparently distinct senses of what it means to be a single whole with respect to arithmetical considerations. And in more generalized contexts, the sense (or senses) in which a singular entity displays a determinable dimension of unity also becomes an issue, or perhaps a series of issues.

It seems fair to conclude this phase of the schematic by noting that with an eye toward crucial metaphysical implications, the description of aspects and dimensions of unity and its function when predicated in ways intended to evoke the substantiality of individual things is theoretically significant. For even if 1 is not a number (as the first element in the sequence of natural numbers), 1 taken in isolation can and often does stand as a designator of a single entity. Either way then, Whitehead's metaphysical concerns seem to apply. Thus, if we take 1 in isolation (whether number or entity-designator), is it subject to any kind of process considerations? If not, then an instance of unity exists, presumably at a fundamental level of functionality, *apart from* process considerations; the resulting reality would certainly appear to be an important element to integrate theoretically as, or at least to situate within, an especially foundational level of Whitehead's metaphysics.

b. If a process of fusion is required to produce 2 (from 1 and 1), does the same process of fusion produce 3 (from 2 and 1, or from 1 and 1 and 1)? As just noted, even if 1 is not a number, 1 as a descriptor certainly appears to generate a dimension of unity; but so do 2 and 3, and it will surely not be denied that 2 and 3 are numbers. Thus, if for simplicity's sake, we assume that 1, 2, and 3 are all numbers and that each is *one* number, then it becomes a question of how to understand the unity of 2 and 3 *qua* number, especially if 2 and 3 are *de facto* numbers and 1 in some sense lies outside the scope of the modes of instantiating unity exhibited by 2 and 3. How can the unity of 2 as one number be thought of with respect to process, the unity of 3 as one number with respect to process, and so on?

c. A brief introductory comparison of geometrical figures (e.g., triangles) with numbers will be instructive. Triangles are identities of a certain definite configuration but also have inherent possibilities of variation. Thus, the lengths of the sides of a triangle (square, etc.) can increase or decrease without affecting the triangle's mathematical type; by contrast, numbers are not open to the same kind of variation with equivalently neutral effects as such variation pertains to the triangle viewed as a mathematical "primitive." Parts of triangles can be changed without having any bearing on that figure's basic nature. Numbers appear to be fundamentally different from geometrical figures in this respect. It seems fair to conclude then that the unity of a number as a mathematical entity differs from the unity of a triangle as a different kind of mathematical entity.

We have concentrated on the implications for mathematics based on the *Phaedo*; later, in the *Republic* and *Philebus*, Plato presents an approach to the reality of numbers that differs in certain fundamental respects. Socrates speaks there of "units" as resident in numbers such as 2, 3, and so on; thus 2 contains two units, 3 contains 3 units, and so on. But how can one number have two constitutive units (or, in a word, parts) and another number have three constitutive units or parts if both numbers are, in themselves, identical with respect to the predication of unity, that is, 2 and 3 are both *one* number? In Chapter 5, we will review the Platonic position on numbers developed with respect to containing units and also lay out some of the most potent criticisms made by Aristotle against the intelligibility of this position. For now, we note by way of possibility that such a specification of units may constitute a type of process. Thus, if 2 in point of metaphysical fact has two elements, 3 has three elements, and so on, then identifying the precise membership of elements in a number can be taken as an instance of a type of unifying process rather than a sheer static enumeration of these elements.

Static Groups

d. Whitehead's characterization of "numerosity" with respect to a group suggests that a group is capable of being assigned a number precisely because it is "static"; if so, then the senses of unity functioning with respect to the group and its members become even more intriguing. A group of seven fish in the water is on the move, swimming this way and that; as a result, this group is certainly not static in the sense of any of its members being immobile. Since, following Whitehead, the group can be enumerated, we must assume that each member of the group is and remains "one" particular entity (in this case, a single fish distinguishable from other fish). What then is "static" with respect to the group understood as such, as a singular entity encompassing a set of individual entities with an assignable number enumerating the extent of the group's membership? Is it simply that a group is a collection of entities greater in number than one? If so, this sense of unity appears unduly exclusionary given that each member of the group is (i) one entity in itself as well as existing, as one entity (ii) apart from the other members of the group.

Also to be noted in this context is the fact that all members of the group may be, but of course need not be, instances of one certain kind (i.e., of fish, given the example at hand). If, however, the group includes fish of more than one type, then the unity of a group so characterized does not appear to be "static" with respect to type determinations. As a minimal concern (and leaving aside type considerations for now), this Whiteheadian-inspired perspective on substantial process indicates that the unity of a single member of a group should be contrasted with the unity of the group as a whole, as such (and regardless of how many individual members may comprise that group—for example, a single "catch" of fish).

This conclusion, that the unity of a group exhibits different characteristics from the unity of any single member of that group, may be taken to illustrate a working (and presumably metaphysically fertile) example of a principle enunciated at *Republic* 525a. Socrates has related to those in attendance a situation when the mind is presented with an apparent contradiction while looking at one and the same finger described as both big and small in relation to adjacent fingers. As a result, Socrates contends that "the mind will be forced to seek a way out of the difficulty, setting thought in motion and asking what unity means. In this way the study of unity would be one of those which convert the soul and lead it to the contemplation of reality."

The reference to "thought in motion" in this context is ancillary to the demarcated sense of process involving particularity and the perception of contradiction (since thought, when thinking, is *always* in motion and thus always in a process condition). The relevant point concerns process with respect to what the mind is thinking *about*, that is, determining senses or dimensions of unity. Unity as a gathering or collecting agency has become diverse: the substantiality of a single element in contrast with the substantiality of a unified (in some sense) group of elements—and, as a result, redirecting the mind toward "the contemplation of reality," those dimensions of the real represented by the Forms, which by their very nature are not subject to generating such contradictory experience.

Procedural

If particular entities exist by themselves, it is worth investigating whether the *process of gathering* entities, or even designated properties of entities, has any effect on the subject matter so gathered. This question presupposes that it is possible to separate and identify the process involved in gathering particulars from the particular entities themselves. Thus, even if the process characterizes the particulars in certain ways (e.g., with Whitehead, whatever processes animate the mathematical operations of addition or multiplication), the results of the process as applied to and marking particular entities (in this case, numbers such as 2) are distinguishable from the process itself, taken as a kind of motion (or, again, fusion). The results of the process are "in" particulars but the process itself represents a different kind of reality. As a note of anticipation: This distinction becomes relevant, if not crucial, to the Platonic position on collection and division initially discussed below (Chapters 3 and 4).

In general terms, we may describe these phenomena as involving *procedural* processes, with the redundancy in terms only apparent in view of the following possibilities (which are not exhaustive given this kind of approach to realities under scrutiny). If a particular singularity has been determined and identified as such, that is, as one particular entity (via a mode of substantial process), does anything happen to the unity of that entity insofar as the entity has been gathered and then embodies dimensions of process in itself? Do these processes have structure, are there distinctions between or perhaps even among types of such structures and, if so, are some types teleological in some respect?

Other prospective possibilities also arise. For example, does fusion of any sort always result when diverse entities are gathered, or are there types of gathering that do not culminate in fused realities? If the latter is the case, then such gathering preserves the unity of some, perhaps all, of the individual entities gathered (in contrast to fusion, during which the original unities are seemingly dissolved or at least significantly altered in structure). But if the former is the case, is it possible to articulate phases of a given type of gathering process that accomplish a certain end, for example, how does 3 come after 2, 4 come after 3, and so on? Thus, are mechanisms of process operative in order for fusion to occur in such a way as to preserve *a certain sequential order*? If so, how would these mechanisms be described in our sample context of numbers with respect to any number greater than the number 1 (assuming, obviously, that 1 is a number)?

These considerations suggest that Whitehead's "unity of issue" (MT, 97– quoted above) is ambiguous in metaphysically significant ways—the phrase can refer either to (i) the result of a process of issue, that is, one number indicating the extent of the membership of a group, or (ii) the process itself by which a group of distinct elements is capable of being identified with one number (as in specifying that a group is constituted as two fish). It may be observed that this possibility reinforces the assumption made above in this section, that is, that the process, discerned as a single process, can be considered apart from whatever number establishes the final "static" unity of the members as grouped.

Textual

The radical character of Whitehead's challenge to the philosopher of mathematics suggests that it would be prudent to spread a wide net, as it were, for approaches to the notion of process which would facilitate the possibility of integrating or involving its elements into the core of theoretical statements pertaining to mathematical realities. So far, the dimensions of process identified in the schematic preamble refer to mathematical entities as such or in relation to one another. But if we take a cue from Whitehead's own procedures, we may discern another approach to the relevance of process in this area which, although entirely heuristic, may prove to be of considerable value in subsequent inquiries. Whitehead juxtaposed Plato's *Timaeus* and Newton's *Scholium*, and in the process of these two texts intersecting with one another, he saw an affinity between a Platonic dialogue and modern physics; he then juxtaposed Platonic approaches to the metaphysical character of numbers with basic tenets of process philosophy and registered apparent flaws in the epochally earlier accounts.

In this procedure, we observe the *textual* process, that is, the recognition that sources developing disparate theoretical perspectives, placed in a kind of interpretive proximity to one another and then examined, may reveal important aspects of positions which, absent such procedural comparison, either may not have been discovered at all or perhaps were noted tangentially but not properly appreciated. This dimension of the process does not contribute any substantial or relational insight as such to the metaphysical matters under analysis, but it does invite the investigator to take seriously the possibility that the thinking of apparently very different philosophers may intersect in places where close scrutiny of shared likenesses (in tandem with the recognition of basic differences) may generate significantly novel modes of conceptual determination. This kind of deep-level interplay, upon due analysis of the combined results, may yield insights into the explanatory coverage and structure of a given metaphysical position.

The reader will observe that we have, in fact, already introduced textual process in the juxtaposition of Plato with both Whitehead's *Process and Reality* and *Modes of Thought*. This comparative survey and subsequent derivation of consequences produced (as illustrated in the schematic preamble section above) a series of heuristic dimensions for determining the relevance of mathematical considerations on the working practices of the metaphysician. The explicit inclusion of "textual process" as one component of the schematic preamble is therefore, in a sense self-referential with respect to the argumentative structure of this book itself (and thus perhaps arguably redundant, at least in this narrowly defined context). Finally, as an anticipatory note, we shall employ textual process again in Chapter 4, that is, juxtaposing an area of Plato's later metaphysics with group theory, a modern discovery in mathematics, and also in Chapter 5, where the confluence of texts to be scrutinized—from Plato and Aristotle—involves additional factors (described at that point). Also worth noting here is that Chapter 8, the conclusion of the book, will explore this procedural possibility and its implications inclusive of metaphysics in general.

Summary and Prospects

Whitehead's responses to Plato's positions on mathematical elements cover distinct areas of speculative thought. And although he sternly criticizes the Platonic approach to the metaphysics of numbers, Whitehead nonetheless introduces a formulation of abstract metaphysical elements—"identities amid diversity"—which, duly developed, will incisively apply to metaphysical strategies Plato pursued in later dialogues. Indeed, if we return to Plato in work done after the *Phaedo*, we will appreciate the extent to which a Whiteheadian review of the sort presented in this chapter can serve as a prelude to establishing more penetrating connections between mathematics and metaphysics.

One key question, as noted earlier, concerns the programmatic status of the identities produced by various philosophical procedures. But the other factor, the diversity of elements from which these identities can be derived, is equally fundamental to this context of inquiry. As we shall see, the attempt to articulate the content of relevant identities and the limits imposed by accompanying diversity will help elicit important systemic connections to a fundamental area in modern mathematics. The varied dimensions of unity and questions related to these approaches, as raised in the schematic preamble, will contribute to stating and justifying these connections.

To summarize: Plato introduced the thematic intersection of mathematics and metaphysics; he also took a strong stand on the effects of metaphysical considerations on the ways in which we conceptualize some of the most basic elements of the mathematical enterprise—figures and numbers. A. N. Whitehead noted these Platonic concerns in mathematics and reacted to them variously as well as vigorously, couching his responses with positions deployed at an especially fundamental level of metaphysical reflection. The principal lesson from this exchange involves realizing that if we approach the two disciplines from a certain interpretive perspective, it becomes evident that metaphysicians and mathematicians work on levels of equivalent generality and employ elements sharing an equally fundamental mode of expression. These connections, if examined in certain broadly specified contexts, will contribute to a deeper understanding of the relevance of modern mathematics to the metaphysical enterprise, that is, recognizing mathematical elements as transcribed into natural language and then, expressed in appropriate linguistic format, doing sustained metaphysical work elucidating the underlying structure of types of reality.

The primary question driving the remainder of this book may therefore be formulated: what can the metaphysician learn about the subject matter of metaphysics by analyzing the procedures and achievements of mathematical inquiry? Chapters 3 and 4 jointly establish a broadened foundation for pursuing this kind of investigation; Part II of this book offers ways to state the results of this high-level comparative endeavor.

Notes

1 Alfred North Whitehead, *Process and Reality,* Corrected ed. (New York: The Free Press, 1978), 94–5. Hereafter, cited as PR with the appropriate page number. The philosophy as "footnotes to Plato" passage appears on 39.

2 F. M. Cornford, *Plato's Cosmology* (New York: The Liberal Arts Press, 1957). Citations include the page number in Cornford as well as the Stephanus number for the text of the *Timaeus*.

3 In the Preface to *Plato's Cosmology*, F. M. Cornford briefly reviews A. E. Taylor's commentary on the *Timaeus* (published in 1928), indicating a tendency which, from Cornford's perspective, distorts Taylor's reading of the dialogue: "the practice of translating Plato's words into the terms of Professor Whitehead's philosophy" (Cornford, *Plato's Cosmology*, ix). Cornford gives examples from Whitehead and Taylor to support this critical reaction (Whitehead's *Process and Reality* appeared in 1929 and Cornford's commentary in 1937). The purpose of using Whitehead in this book is very different from the insinuated doctrinal intimacy which Cornford accuses Taylor of perpetrating. Whitehead outlines a subtle but illuminating thematic connection between moments in the *Timaeus* and certain strains of modern science; we will use a similar connection to examine the approach to the metaphysics of numbers developed in the *Phaedo* (and codified in an important basic element in modern abstract algebra) in relation to metaphysical positions advanced in later dialogues. If the examination is successful, the reader will recognize a series of affinities between this phase of Platonic metaphysics and elements of modern abstract algebra that illuminate important aspects of the former. But the narrative strategy of including Whitehead only gives point to dimensions of Platonic methodology and metaphysics that become more apparent when we embrace and articulate the implications of juxtaposing Plato and modern mathematics. We are using Whitehead's example to help establish a critical response to important elements in later Platonic thought—bluntly put, we are not reducing Plato to the status of a proto-Whiteheadian.

4 Alfred North Whitehead, *Modes of Thought* (New York: Free Press, 1968). Hereafter, cited in text as MT with the appropriate page number.

3

Plato and the Notion of Grouping

Collection and Division

Whitehead's divergent readings of Plato presuppose a significant uniform assumption—history as the sheer passing of time has no effect on the discernible realities selected for comparative and critical analysis. The extensive period separating Plato from modernity is not significant with respect to identifying the relevant content of what has been asserted over this stretch of time. Informative comparisons in forward directions (the Platonic use of triangles as metaphysically anticipatory formative elements) and in backward directions (the critique of unduly static formalisms posited as foundational for the metaphysics of numbers) are sanctioned solely by the mathematical and metaphysical character of the constitutive elements under scrutiny, not by when in history the elements were initially proposed and the extent of time separating these origins from comparative analyses. The emphasis, again, is on content, what has been asserted and what can be informatively related to what has been asserted. This rudimentary watchword, procedural (and perhaps perfunctory when working within abstract philosophical contexts), may nonetheless help orient a measure of forward-looking attention toward the following reflections on Platonic practice. In that regard, this chapter will contribute essential foundational material for the epochal identities (thus applying Whitehead's *Modes of Thought* claim) which ground the speculative analyses advanced later in Part II.

Grouping: Collection As Methodology—Introductory Comments

According to Socrates' argument in the *Phaedo*, grouping instances of the number 1 does not *cause* the number 2 (although 1 + 1 *equals* 2 from the standpoint of arithmetical protocol). It might appear then as if Socrates is in principle against grouping elements when fundamental metaphysical issues are under consideration on the grounds that, as a procedural ploy, grouping imposes realities localized within certain arbitrary and limiting boundaries in order to generate a makeshift unity as an explanatory causal account of a completely different reality. Socrates has denied that the process of grouping elements 1 and 1 is an appropriate step in accounting for the metaphysical origin and proper cause of the number 2. It would, however, be premature to conclude that for Plato, *grouping in any sense* is disqualified as an essential phase in a foundational metaphysical process. Indeed, it would be a mistake.

Appropriate methods for pursuing philosophy were a recurring theme in many Platonic dialogues. One method—typically referred to as "collection and division"—appeared in the *Phaedrus*, with later installments of this methodological procedure animating the *Statesman* and *Philebus*.[1] Our initial concern is with collection as a phase in philosophical method, and we shall examine specimen passages from these dialogues to clarify this process. The analysis of number in the *Phaedo* yielded consequences signaling that we must take special care when considering individual entities for metaphysical purposes. It will become evident, however, that we should not jettison the notion of grouping altogether—in fact, gathering particular entities into one uniform group becomes, and, properly characterized, stands as a methodological necessity.

Consider this passage from the *Phaedrus* (265d)[2]: "we bring a dispersed plurality under a single form, seeing it all together—the purpose being to define so-and-so, and thus to make plain whatever may be chosen as the topic of exposition." Socrates' interest at this juncture of the *Phaedrus* concerns distinguishing types of *eros*. In his reflective analysis of a speech on *eros* he had presented to Phaedrus earlier in their conversation outside the walls of Athens (while he was in a state of inspiration, according to his own testimony), Socrates "casually" introduced the method of collection and division; in fact, the speech itself had been contoured according to that very method (265c-d). We note then that bringing "a dispersed plurality" so that members of this plurality are "all together" clearly represents a process of gathering. Furthermore, the gathering has a specific purpose—to place an interested party in a position to discern something real and present to the individual entities so gathered which will serve as the appropriate metaphysical ground for determining a definition of the reality under philosophical scrutiny.

Particulars are collected by virtue of recognizing that certain items possess shared, similar properties. Once these items have been gathered into one group in order to facilitate reflective examination, the philosopher then divides the collected particulars in such a way as to elicit the appropriate Form (or, perhaps, Forms) that underlie—or, in the terminology of the *Phaedo*, cause—these particulars. Although the statement of collection and division introduced at this point is spare (we will supplement the text quoted with related passages from later dialogues), it is useful to note the breadth of relevant aspects of collection and division stated and implied by the *Phaedrus* announcement of a new procedure for securing metaphysical ends.

A. Socrates does not specify to what extent the constitutive members of the plurality are or have been "dispersed," but we may plausibly assume that the field of inquiry encompasses a number of individual entities; it is not possible to read off substantive metaphysical results from a single entity.
B. The entities to be gathered are "spread out" over a certain spatial and/or temporal region; if so, then the investigator must be alert to the practical requirements needed to gather all relevant items in order to fulfill the ultimate metaphysical purpose of such gathering.
C. We are not told whether the process of collection requires a specific number of items or what could be considered a representative sampling of the elements under scrutiny: "plurality" is indeterminate in this regard.
D. Can the collection proceed by bunching particulars (e.g., a "handful" or two of items) or is the process of gathering to proceed one by one, each particular examined by itself as a procedural prerequisite for inclusion

in the group? If the latter, adding particulars individually, then it is important to appreciate that from this perspective of a methodology geared for metaphysical purposes, one as a number uniquely identifies each particular to be added as well as distinguishes this particular from that particular, and so on. The number one thus functions in a dual capacity—it enumerates each entity gathered (if it is the case that the total number of entities gathered is a factor in a successful application of the method) and also refers to each particular as a unit entity, that is, an individual totality (presumably relevant to the process of division as it seeks to identify formal structures characterizing the set of particular units gathered).

E. Insofar as the collection phase illustrates a process of grouping, the only requirement for this process stated in the *Phaedrus* is that the particulars collected should display similarities between and among the items so grouped. Presumably then this generously open-ended stipulation allows as many or few elements as needed (whether grouped individually or in bunches) in order to generate a kind of critical mass suitable for the identification of relevantly causal Forms.

Finally, we may recall Whitehead's proclamation that the purpose of metaphysical inquiry is to seek "the discovery of identities amid diversity" (MT, 98) and that this specification tersely but neatly announces a significant function of the Platonic method of collection and division. How to approach the relevant identities as they emerge from within contexts marked by diversity with due appreciation for whatever implications the investigator derives from these instances of generalization constitutes crucial dimensions of inquiry and will be discussed later in the book.

For now, we may conclude that not only does Plato have nothing against grouping as such, but the process of grouping particular elements, in point of metaphysical fact, has become a necessary methodological fixture in the process of establishing formal features animating what is real. The primary restrictions explicitly introduced in the *Phaedrus* account are, first, however many individual instances are gathered from an original condition of dispersal into some kind of grouped commonality, each item must bear some similarity to all the others within this group. And second, inquiry is guided by the working assumption that such grouping collectively encompasses a discernible unity, a kind of class common to and derived from the individual instances so grouped (regardless of the number of these instances). It may be inferred then that the particulars gathered would not be so dissimilar among themselves that attempted specification of commonalities which might appear to characterize a given assembly of entities would fail simply by dint of their sheer multiplicity and diversity of content.

Numbers and Processes of Counting

The methodology of collection and division is intended to identify Forms derived as constituents encompassing a range of particular entities. Numbers will presumably be included in this range, but the method will apply, again presumably, to virtually any kind of particular. The wisdom aimed at in the pursuit of philosophy encompasses everything that exists. It will nonetheless be convenient for purposes of explication to consider numbers as, initially, the primary metaphysical focus of interest. Numbers are advantageous in this regard since they are inherently abstract realities that can be ascribed to or predicated of virtually any kind of entity, whether real or imaginary. Furthermore, in addition to their own dimensions of reality, numbers may be considered placeholders for more complex kinds of particulars, but particulars are also subject to the same types of theoretical examination. The investigations to follow will determine whether characteristics (a) applicable to numbers when grouped with one another as a set of singular entities will also (b) pertain to the entities counted by those numbers. If so, then principles that apply to numbers construed in a group context will also apply to metaphysical situations of referential diversity and complexity.

Generally speaking, the "modes of counting" in the title of this section refers to reflections on the notion of unity as its complexity is gradually revealed during the conjoined processes of counting and identification. Since a point of departure is required, we will provisionally view numbers as particular entities, unities, or wholes unto themselves (leaving open for the moment any sense that numbers are wholes of parts) and existing in certain relationships to other numbers. Recall also that a group of particulars instantiating the collection phase of Platonic methodology will encompass an unspecified number of elements. This lack of specification refers to the total number of elements constituting the group, whatever that totality may be. However, regardless of whether few or many items populate that plurality, they will presumably be denumerable (not a loose or informal "some" or "a large number"). This constituent feature of a group—its enumeration—may be isolated and then considered as a separate factor integral to the overall structure of the group. In other words, we will assume that the investigator in search of identifying Forms is gathering not only *individual realities* incorporating elements that instantiate determinable content, that is, pertaining to whatever is currently under philosophical scrutiny, but also *the number of* realities displaying those elements, realities that, collectively, constitute the membership of a given group.

To enumerate members of a group is to count them. The Greeks used numbers to count things, as we do now. An inventory of house supplies done, for example, by a contemporary of Plato may indicate one container of wine on hand—or perhaps none; that is, no single container of wine on the shelf (time for a trip to the agora, perhaps. . .). Let us suppose that no containers of wine on the shelf are unacceptable, but one container might be sufficient. Now juxtapose 1 and 0 to form a single group of numbers and limit the grouping to exclude other possible numerical entrants (two containers of wine are potentially an invitation to excess and thus do not represent, for this household, a live option). Those are our numbers, derived from reflection on a specific and quite practical state of affairs in household management.

At this juncture, the argumentative strategy, broadly stated, begins with mathematics and then derives consequences relevant to metaphysical inquiry. To that end, we propose to examine the structure of a collection of items containing two numbers—0 and 1. However, an immediate objection arises aimed at this example. For if we assume that the Euclidean definition of a number (*Elements* VII), "a multitude of units," was operational in the time of Plato and Aristotle, then neither 0 nor 1 is a number, although for different reasons.[3]

So formulated, this issue is primarily historical—what did the Greeks of Plato's time construe as numbers? According to Anders Wedberg, in *Plato's Philosophy of Mathematics*, "the Greeks, including Plato, were not infallibly consistent on this point."[4] If this lack of consistency among Greeks regarding the status of 1 and 0 as numbers is indeed the case, as Anders maintains, then even from a strict historical perspective, we are justified in reexamining 1 and 0 with respect to their possible function as numbers from the standpoint of a contemporary of Plato and indeed from Plato himself. The viability of this context of inquiry is important because theoretical interests might range beyond implications restricted to common usage. In other words, even if the Greeks consciously and consistently rejected both 1 and 0 as numbers in their practical *counting*, a *theory of numbers* might be formulated that could incorporate in its scope either or both 1 and 0.

If we limit inquiry at this point to the more prevalent and metaphysically dominant of these two numbers, we will focus on the first of the two. What then can be said, even if only in an introductory manner, about 1 as a number?

The Number 1: Relational and Substantial Considerations

Although unity functions as a blanket term of admittedly high generality, it nonetheless can convey seminal information. The number one is an instance of unity. Furthermore, the introductory comments on collection (in the first section of this chapter) outlined the diverse functionality of the number one—in (a) counting as well as (b) the process of referring to a particular insofar as its mode of existence is characterized by singularity. Given the importance of one as an instance of unity construed in this multifaceted way, it is crucial to determine, as precisely as possible, how the number one should be understood in these settings. To achieve a measure of clarity in this regard becomes all the more important given the philosophical interest in the collection of particular, that is, singular items when this process stands as a component of a methodology intended to secure fundamental metaphysical ends.

We explore this question initially by discussing what Plato actually wrote in this area. First, a prefatory note on general context: in Chapter 1, we introduced the account of the generation of numbers in the *Phaedo* in order to initiate the speculative position argued in this book. And although the reflections at hand are derived from the *Phaedo* as well as other Platonic texts (and, indeed, can stand as viable readings of those texts), their primary purpose is to offer a theoretical foundation for more expansive approaches to the interactions between mathematics and metaphysics than are stated in or directly implied by the *Phaedo* taken in isolation. The analyses concerning whether one is a number are therefore not solely intended to deliver evidence on just this issue when pursued in a strictly Platonic context. Rather, the discussions of each passage cited also develop various senses of unity. Furthermore, these discussions illustrate, in what we may hope are fruitful applications, *textual process* and *substantial process* (subdivisions of the Whiteheadian notion of process introduced toward the conclusion of Chapter 2).

1. In the opening line of the *Timaeus*, Socrates is addressing his audience: "ἐις, δύο, τρεῖς. . . —but where, my dear Timaeus, is the fourth of those guests of yesterday who were to entertain me today?" (Cornford, 9/17a) Socrates, counting the house as it were, wonders why the audience for the day's conversation is one less than expected. The first three words of the *Timaeus* are the first three numbers in a typical counting sequence.[5] Why Plato began his grandly depicted cosmology with this seemingly humble arithmetical ploy is perhaps worth some reflection given the interpretive assumption that the very beginnings of dialogues can carry significance for what follows. But for present purposes, we note only that in counting from one to four, Socrates enumerates the expected number of conversational participants—presumably a complete set of these individuals.
2. In Book 7 of the *Republic*, Socrates is developing his program of studies for a future philosopher-king. As a prerequisite, Socrates identifies "this common thing that all arts and forms of thought and all sciences employ, and which is among the first things that everybody must learn." What is it? "This trifling matter . . . of distinguishing one and two and three. I mean, in sum, number and calculation" (522c5-8).

   The ability to do what Socrates does in the *Timaeus*, count how many people are present, requires a grasp of numbers, and in the *Republic*,

Socrates asserts that the ability to distinguish between and among 1, 2, and 3 is "trifling" (a sequence of 1 through 4 would be only slightly less trifling). Note the distinction between "number" and "calculation." We must learn our numbers, but we must also learn how to calculate with them. Socrates says that in both areas, exercises of the sort mentioned are trifling. Thus, to know about the number one is immediately obvious (also to know that 1 + 1 = 2). However, as we have been arguing, correct calculation is not equivalent to the desired level of metaphysical analysis. Numbers can be taken individually, one by one, but they also occupy a sequence involving multiplicity. And we note that this aspect of numbers, if viewed again from a Whiteheadian perspective, necessarily injects process elements into a proper metaphysical understanding of the full reality of numbers.

3. The Socrates of the *Phaedo* asserted (97b): "I no longer believe that I know by this method even *how one is generated* [my italics] or, in a word, how anything is generated or is destroyed or exists." The key qualification is, of course, "by this method," the addition of numbers by arithmetical convention or the partitioning of lines. Socrates' concern is to account causally for the existence of the number one; this inquiry is completely different from the ability to employ the number one as an element in counting, as Socrates effortlessly demonstrated in the *Timaeus*. Thus, it is "trifling" to distinguish one, two, and three when we are counting things; however, it is not at all trifling to account for the existence of 1, 2, and 3 as unique numbers. The correct approach requires serious and sustained metaphysical inquiry (procedurally attested to by the Socratic period of self-imposed silence mentioned at *Phaedo* 95e).
4. The existence of the number one is, in fact established by the theory of Forms. Socrates says so in the *Phaedo*. As we saw in the opening chapter above, "whatever is to be one must participate in unity"; thus, the Form identified as unity or oneness causes any single thing to exist as one individual, including that one individual which is the number one itself. What is asserted as a matter of theoretical principle in the *Phaedo* is underscored in the *Statesman* (258d)[6]—if we "consider the science of number and certain other sciences closely akin to it," we will

> understand that they are not concerned "with any form of practical activity" but rather with "yielding us knowledge." To know the existence of numbers in the most basic sense, including the number one, is to appreciate the necessity to know the Forms; this level of cognitive awareness transcends, in its subject matter, the level of practical activity that characterizes, if not monopolizes most human interaction with numbers—that is, counting things and basic exercises in calculation.[7]

For Plato, it is vital to distinguish between one as a number used in counting as well as calculation, and a theoretical interpretation of the number one. Human beings, as practitioners of calculation, readily attach the number one to a particular thing (as long as this entity is experienced as a determinate given) and then, if interest and circumstances obtain, continue with the count from 1 to a number "higher" than 1. However, empirical recognition and connection to a series of given integers differ fundamentally from a theoretical explanation demonstrating how those integers exist in the first place. It is essential, then, to separate the counting function of number and its obvious utility in many practical contexts from the formal existence of number.[8]

To summarize—Platonic texts support the following: 1 is a number; 1 is the originating member of a sequence of numbers; these elements as well as the sequence itself are all "easily" recognized—by contrast, the existence of 1 as a number is problematic (even if its practicality is manifest); however, the theory of Forms accounts for the existence of the number 1. Thus, from a strict philosophical perspective, the reality of the number one in all its diverse computational functions ultimately depends on explaining, at an appropriate level of metaphysical generality, how that number exists.

Grouping: Structure, Membership, Perspective

The above summary provides textual evidence that at least for Plato, one can indeed be construed as a number. This summary can be taken to support Wedberg's generalization that the Greeks were "not consistent" about the denial of one as a number. With respect to the other element in our proposed set of elements—0—we need not be so cautious. Reginald Allen comments: "By modern mathematical standards, the Greek concept of number was impoverished" (RA, 263); after all, they lacked awareness not only of 0 as a number but also the existence of negative numbers. But at this point—and with the dominance of the "binary code" in the world of computer technology (which means, of course, in the world) functioning at full force—we must separate the metaphysical interests to be pursued in this book from the parallel concerns of ancient Greeks and their philosophers, especially with respect to their habitual (and "impoverished") understanding of numbers. Thus, for modern mathematics, the use of the number 0 and the number 1 elicits no theoretical reservations. Therefore, the relevant point from this updated perspective turns on whether the possibility of grouping 0 and 1 makes sense mathematically (even if the Greeks themselves did not recognize 0 as a number and even if some of their philosophers—for example, Aristotle—offered arguments supporting the position that one is not a number).[9]

In fact, this group—0 and 1—is very important. If the operations between 0 and 1 are defined as follows:

$0 + 0 = 0$
$0 + 1 = 1$
$1 + 0 = 1$
$1 + 1 = 0$

Then this configuration is the basis for the binary code, initially formulated in the late seventeenth century by Leibniz and which, in the twentieth century, became the foundation of computer science. This formulation took mathematics beyond the realm of the known numbers: natural numbers, integers, rational numbers (fractions), and real numbers (which include the "irrational" numbers, for example, the square root of 2)—instances of numbers typically dealt with from a practical standpoint. Removing 1 + 1 as a conceptually necessary component of a given group allows (a) the addition event to be kept apart from (b) the formal circumstances that control the determinate outcome of that event. Thus, when Socrates in the *Phaedo* separates 1 + 1 from 2, he opens up the possibility that 1 + 1 represents an operation that might produce something other than the "expected" result when counting with numbers in the traditional manner. In fact, this approach to an alternate grouping, combined with the operation of addition, leads to: $1 + 1 = 0$.

The contrast, at least at first glance, between $1 + 1 = 0$ and $1 + 1 = 2$ is dramatic. Furthermore, assume that the collection (0, 1) is rejected—an approach which, we have argued, would be primarily based on an uneven, mixed reading of Greek thought and certainly open to reconsideration. The crucial point given the framework of this book is that *both* arithmetical propositions are true (or, if one prefers, viable) if we allow the same number—1—to operate within different mathematical structures. Thus, the grouping structure of the two-element set (0, 1) is necessary as a factor underlying the validity of $1 + 1 = 0$. For if such an aggregation is closed and the addition of its sum total of individual elements is performed, then 0 is the only possible result. It is true therefore that if the number one is separated from its function as an indicator of counting perceptibly distinct particulars, then its unity as a determinate number and its membership in a certain defined collection allows it to be employed in an operation the results of which look paradoxical. It should be noted, however, that this disorienting impression occurs only because we are much more accustomed, given the practical demands of computation in relation to individual entities, to counting in sequence from 1 to 2 to 3, and so on (where $1 + 1 = 2$) than we are—unless we make our living as a computer programmer—counting from 0 to 1, then from 1 to 0, and so forth.[10]

Methodology and the Identification of Forms: A Preview

Even if the Greeks themselves had no obvious working sense of 0 as a number, we have argued that this omission coupled with the ahistorical character of mathematical entities and the present acknowledgment of the legitimate numerical status of 0 will permit an examination of a group—the binary code—with membership of two distinct elements, 0 and 1. The strategy underlying the formation of the binary code reveals two basic phases: (a) the process of gathering which leads to producing a collection of items or entities each identifiably distinct from one another; (b) the deployment of these items as members of one specific instance of grouping in a manner facilitating the possibility of producing some sort of result. In this case, the result is not only possible but palpable—the technology that has begotten the machine on which many texts are written (such as the text the reader is presently examining!) depends to a considerable extent, at least mathematically, on a binary code comprising just two elements.

The reformulated system of counting introduced in this section, if accepted as a feasible development of the Platonic agenda on the nature of numbers in the *Phaedo*, provides a theoretical backdrop for a redirected approach to grouping or collecting elements for specific mathematical purposes. With this newly defined sense of aggregation in hand, we will be in a position to assess critically a related but purely philosophical approach to determining fundamental dimensions of metaphysical inquiry. The Platonic description of collection represents a kind of gathering or grouping, the initial phase of the method of collection and division. Two dimensions of this method must be analyzed—describing the phases of the method as fully as possible; clarifying the description of the kinds of realities that the method intends to produce. This clarification should be articulated in ways that reflect description of these realities (given the generalized statements of the method). Now consider the two contexts at hand: (a) mathematicians establishing a group as a type of unity and how they populate the membership of this group, and (b) Plato conceiving the process of collection—also a type of unity when a given instance of collection has been completed—and how an instance of such unity encompasses particular items metaphysically constituted according to formal specifications.

Each configuration displays unity; however, the constitution of unity applicable to a collection of only two members differs considerably from the unity of the individual numbers appearing as members of such a gathering (differences of unity also distinguish the results of basic arithmetical operations). We suggest as a provisional inference that the same considerations concerning variant senses of unity will hold when the gathering in question incorporates the Platonic doctrine of collection as the initial phase of collection and division. Furthermore, we may assert as a speculative hypothesis which will guide inquiry in this book that collection and division are *methods with structures common to both metaphysics and mathematics.* If so, then if this method is taken seriously as one way to determine the structure of reality, the claim is that the metaphysician should appreciate how the application of such a method might benefit from perspectives provided by related but distinct forms of theoretical investigation—that is, mathematics, in particular a certain area of modern mathematical theorizing.

The cogency of this direction of inquiry, especially in the reflective contexts to be developed later in Part II of this book, presupposes recognition of at least three distinct senses of unity, all of which are expressible in shorthand form, as it were, by the number one: first, as an indicator for counting particular entities; second, as a referential sign establishing the identity of each single item counted—these two senses of unity separate the straightforward denumerable aspect of the process of counting from the substantial unity (the nature and specification of which can vary widely) of what is counted; third, as a referential sign for a single group or collection (with either a determinate or an indeterminate number of members) along with the understanding that a rationale is available to justify why this kind of configuration has been assigned to this set of individuals (e.g., again, the utility of the binary code). Establishing the collection as itself a kind of unity allows the structure of a group to be investigated and analyzed in terms of properties accruing to a distinctive kind of unity.

The insights provided by a focused investigation of a certain type of group, especially with respect to their significance in relation to metaphysical inquiry, will be identified and discussed in the next chapter. The results of these analyses will conclude the foundational material required to ground the speculative positions introduced and justified in Part II.

Notes

1 The secondary literature on collection and division in Plato is extensive. The texts and discussions considered here are sufficient to bring out mathematical dimensions in Plato's thought as they pertain to collection and division—a primary objective of this book.

2 The translation is by R. Hackforth, *Phaedrus*, in *The Collected Dialogues of Plato*, ed. Edith Hamilton and Huntington Cairns (New York: Pantheon Books, 1961).

3 R. Allen appeals to Euclid for this definition of number and concludes that 0 and 1 are not numbers for the Greeks (Allen, *Plato's Parmenides*, 262–63). D. Roochnik quotes in this context Jacob Klein's "a definite number of definite things" and then says: "For this reason, 0 and 1 are *not* [italics in text] *arithmoi* ["numbers" in transliterated Greek]" (Roochnik, "Counting on Number," 546). See Jacob Klein, *Greek Mathematical Thought and the Origin of Algebra*, trans. Eva Braun (Cambridge, MA: The M.I.T. Press, 1968), 46 ("*definite number of definite things*" is italicized in Klein's text). Our hypothetical group of elements will be rejected if 0 and 1 are not numbers according to this view of Attic counting. In the Introduction to *Principia Mathematica*, Whitehead and Russell consider the question of determining the status of 1 as a number: "The proposition 'a whale is big' represents language at its best, giving terse expression to a complicated fact; while the true analysis of 'one is a number' leads, in language, to an intolerable prolixity." We may hope that attempts to state in language useful metaphysical consequences derivable from sustained reflection on precisely this question in the philosophy of mathematics will not strike the reader as little more than an exercise in "intolerable prolixity." See Alfred North Whitehead and Bertrand Russell, *Principia Mathematica* to *56, (Cambridge: At The University Press, 1962), 2. Also, as an aside, "language *at its best*" [my italics] when used, for example, in a proposition representing the size of the whale appeals to the meaningful predication of the superlative degree of the adjective "good"—this basic grammatical distinction plays an important role in the speculative analysis of the good presented later in this book.

4 The Wedberg passage is quoted by Roochnik, "Counting on Number," 546. See Anders Wedberg, *Plato's Philosophy of Mathematics* (Stockholm: Almqvist & Wiksell, 1955), 23. To support this counter-position, Wedberg cites two passages from the dialogues—*Laws* 818c and *Hippias Major* 302a—citations which Wedberg does not develop to any degree. See also Julia Annas, who asserts that Plato "reflects this indecision" regarding the status of one as a number. *Aristotle's Metaphysics M and N.*, trans. Julia Annas (Oxford: Oxford University Press, 1976), 11. Annas' section "Plato's Philosophy of Mathematics," 3–26, provides useful background for this and other problematic areas on this subject. See also: A. E. Taylor, "Forms and Numbers: A Study in Platonic Metaphysics," *Mind* 35 (1926): 419–40; Richard D. Mohr, "The Number Theory in Plato's *Republic VII*

and *Philebus*," *Isis* 72, no. 4 (1981): 620–7; Ivor Bulmer-Thomas, "Plato's Theory of Numbers," *Classical Quarterly* 33, no. 2 (1983): 375–84; Florin, George Callan Florin, "Numbers, Ontologically Speaking: Plato on Numerosity," in *Numbers and Numeracy in the Greek Polis*, ed. Robert Sing (Leiden, Netherlands: Brill, 2021): 219–36.

5 The numbers one, two, and three are repeated at *Laws* 818c in the context of the Athenian giving examples of items the awareness of which illustrates a minimal level of education.

6 Translation (slightly modified) by J. B. Skemp, *Statesman*, in *The Collected Dialogues of Plato*, ed. Edith Hamilton and Huntington Cairns (New York: Pantheon Books, 1961). Subsequent passages from the *Statesman* are also from this translation. For commentary discussing this passage in the context of the complex structure of the *Statesman* as a whole, see David A. White, *Myth, Metaphysics and Dialectic in Plato's* Statesman (Aldershot: Ashgate [Routledge, 2016], 2007), esp. 89–94.

7 Knowledge of the Forms, especially those relevant to the generation of numbers, includes recognition of the fact that since each Form is one, the Forms generating each number are also one. For what has become a classic statement of this feature of Platonic metaphysics, see the account of the mathematical setting of the Forms in: John Cook Wilson, "On the Platonist Doctrine of the ἀσύμβλητοι ἀριθμοί," *The Classical Review* 18 (1904): 247–60. Scholars as diverse in their approaches to Plato as Reginald Allen and Jacob Klein cite this source with approval (Allen, *Plato's Parmenides*, 270–1; Klein, *Greek Mathematical Thought*, 62).

8 R. Allen asserts that "though he [Plato] nowhere explicitly states that one is not a number," this stance is "implied by" *Phaedo* 104a-b. Allen adds in clarification: "This does not mean, of course, that one was not used in mathematical operations: 6 + 1 = 7 is taken as the addition not of a number to a number, but of a unit to a number" (Allen, *Plato's Parmenides*, 263). But it is surely difficult to discern any distinction between 1 as a unit and 1 as a number in the operation represented by 6 + 1 = 7; if so, if 1 should be understood as a unit of measurement, then how is 1 in the operation 6 + 1 = 7 doing anything pertaining to measurement? From the standpoint of basic arithmetic, it is simply the number 1 being added to the number 6. Second, if 1 is, by itself, solely a unit of measurement, then how can it be used to count existing things when there is only one thing there to be counted (such as the lone container of wine mentioned in the example above)? Again, this kind of pragmatic—and very

common—usage reinforces the reality of 1 as a number. Finally, *Phaedo* 101c-d (discussed in Chapter 1) indicates the difference between the Form unity and any particular thing which is one—including the number 1 itself, which the Form unity causes to exist. Thus before 1 can function as a unit of measurement with respect to other numbers and before it can become an element in a calculation, the 1 must exist as a number—participation in the Form unity accomplishes its generation. In sum, the series of passages discussed provides strong evidence against Allen's position (shared by others) and supports the conclusion that for Plato, 1 is a number; indeed, acceptance of the numerical existence of 1 is the reason why Plato (*per* Allen) "nowhere explicitly states that one is not a number." See also Julia Annas' discussion of Greek notions of counting and calculation (JA, 5–6). For more general discussion of Plato and mathematics, see: Harold Cherniss, "Plato as Mathematician," *The Review of Metaphysics* 4 (1951): 395–425; P. Pritchard, *Plato's Philosophy of Mathematics* (Sankt Augustin: Academia Verlag, 1995); Myles Burnyeat, "Platonism and Mathematics: A Prelude to Discussion," in *Explorations in Ancient and Modern Philosophy*. Online (Cambridge University Press, 2012), 213–40 and the work by Anders Wedberg cited in Note 4.

9 The Aristotelian approach to determining the status of 1 as a number will be discussed in Chapters 5 and 6, in the context of developing the main argument of this book.

10 It should be noted that the principles underlying quantum computers, as well as, for example, contextual variations in the analysis of ordinary language, employ non-Boolean logic, which does not recognize the law of excluded middle.

4

Mathematics and Grouping—Collection and Division Revisited

The cognitive interest in identifying how a given group of particulars will yield a Form (or set of Forms) provides the initial impetus for collecting particulars. According to the *Phaedrus*, the philosophically motivated student can, as an initial step in an organized and systematic process, gather a certain number of particular items. Once so gathered, the investigator then groups them in such a way that the resulting proximity of distinct but similar particulars collectively engenders a unified connection (or, perhaps, connections) leading to the possibility of identifying another, more abstract, "higher" dimension of reality—Forms. These connections become available for recognition and subsequent articulation only after the investigator has presumably practiced careful, reflective apprehension of the group's membership and thereby become receptive to identifying the formal structures permeating the group of particulars under scrutiny.

The Notion of Grouping in Algebra

Assume that the particulars we are investigating are numbers. In a strict mathematical context, grouping just two elements, 1 and 0, produces an association of elements which, when posited as a unity, achieves a self-subsistence sufficient to allow the process of standard arithmetical operations to occur (although producing answers that are unexpected and appear initially counterintuitive). However, once we are accustomed to confronting a certain kind of unity composed in a certain way (as do computer programmers), an association or grouping of elements yielding these results assumes a dimension of initial reasonableness and then, indeed, necessity.

When numbers are juxtaposed according to a certain principle of grouping, we recall that 1 + 1 does not equal 2, but 0. The implication, generally stated, is that how numbers are grouped will significantly affect the results of a common operation when that operation is applied to elements formed as sets based on different principles of grouping. In this case, the numbers scrutinized are identical, 1 and 1. Grouping these numbers in one way and in conjunction with a basic arithmetical operation will produce 2; however, when the same numbers are grouped according to a different formal arrangement, the same operation produces different results. Are there principles underlying the generation of such distinct groups? If so, can they be investigated and then, perhaps, applied to philosophical ends?

The passage in the *Phaedrus* announcing collection and division (quoted in Chapter 3) is formulated with definite results in view. And although our working description of grouping coheres with a decidedly metaphysical agenda, it now includes a mode of grouping characterized by more limited and abstract content—numbers—than that comprising particular things or phenomena, the locus of Socrates' field of inquiry in the *Phaedrus*. This transition, from gathering things (understood in the broadest entitative sense) to gathering just groups of numbers, allows the context of inquiry to assume a more limited focus. In fact, philosophers with an interest in examining foundational perspectives in mathematics, especially with equations as they characterize algebra, will be familiar with *Abelian groups*, an approach to mathematical elements which, we will soon see, is similar in structure to a methodology described and practiced by Socrates in various later dialogues. A group in this mathematical sense is a set of elements that satisfies certain stipulated conditions. These elements may or may not be numbers (although we will use numbers by way of illustration at this point in keeping with the examples discussed by Socrates in the *Phaedo*).

Abelian groups were first formulated by the Danish mathematician Niels Henrik Abel (1802–29), who died at twenty-six from tuberculosis. Abel and Evariste Galois (1811–32—Galois died at an even younger age than Abel from wounds incurred in a duel) established the principles that formed the foundation of the area in abstract algebra known as group theory. Abelian groups in particular have become very powerful elements in modern mathematics—for example, in fields, rings, and modules. For mathematicians, Abelian groups represent one of the simplest structures available to prove interesting things. Thus, rudimentary algebra solves equations with variables of the second power by using the quadratic formula. What about equations with powers higher than 2? Applying Abelian groups can show that it is possible to solve cubic and quartic equations but impossible to solve quintic equations in radicals. This is only one instance of the theoretical power inherent in the application of an Abelian group to concrete mathematical situations.

As noted, numbers (i.e., integers) can constitute the set of elements comprising an Abelian group. And since numbers are the focus of our inquiry, we can illustrate the requisite conditions for an Abelian group using numbers to exemplify content.[1]

Closure: the set of elements in an Abelian group must include an operation—addition, for example. Closure means that if any two elements are combined (added, in this case), the result must also be a member of the set. Thus, adding any two integers results in another integer. The number of elements in a set can range from two to limitlessly large for non-trivial groups.

Associativity—for all individual elements a, b, and c in the set, the equation (a + b) + c = a + (b + c) holds.

Identity element—there is an element, e, in the set such that e + a (i.e., any element in the set) = a + e = a holds. For the integers, our sample set, e is 0. Thus, 0 + 2 = 2 + 0 = 2.

Inverse element—for each element in the set, there exists an element, b, in the set such that a + b = b + a = e. If a is 2, then the inverse element of 2 is −2. Thus, 2 + −2 = -2 + 2 = 0. It will be noted that 0 is the identity element for this group.

Commutativity—for all elements of the set, a + b = b + a for any two elements. Thus, 2 + 1 = 1 + 2. Commutativity holds when the operation is addition but does not hold when the operation is subtraction; thus 2—1 ≠ 1—2. The latter sets are referred to as "non-Abelian groups." This distinction regarding commutativity does not by itself affect the symmetry and relevance between Abelian groups and Platonic metaphysics (which we will develop in an introductory way below). Two supplemental points: first, associativity and commutativity, as expressed in the above examples, will be familiar to students of elementary symbolic logic; second, the relevance to metaphysical concerns of the identity element and inverse element conditions (with their pivotal use of 0) will be outlined below.

Plato and Abelian Groups: Introductory Note

For the later Plato, the method of collection and division concentrates attention on (a) individual particulars with respect to the recognition of Forms animating those elements when they are taken collectively, or grouped, and (b) Forms with respect to the blending of individual Forms within a bounded set of distinct particulars. On this level of analysis, philosophical attention is aimed at articulating formal realities insofar as they characterize concrete content. We will proceed at this point by separating (a) the denumerability factor regarding the elements constituting the subjects of metaphysical analysis from (b) the individual realities as such, whatever these elements might be. Now consider the resulting numbers derived from (a) as a sphere of reality in its own right. The above five-part outline states a set of conditions which, upon examination, illustrate that *collection and division incorporate all major defining conditions of an Abelian group, although not all in the same way.* This structural parallel will allow us to show that conditions defining an Abelian group contribute substantially to clarifying a variety of procedural as well as metaphysical considerations when these conditions are juxtaposed with a series of seminal formulations of the method of collection and division as stated in several later dialogues.

To summarize: the approach to numbers in the *Phaedo*, if combined with the application of grouping as a process equivalent in overall function to the collection phase of collection and division, sanctions an incisive interpretive conclusion: just as the Platonic perspective on triangles anticipated certain essential features of modern physics (as Whitehead recognized), so the Platonic approach to numbers and the determination of related class notions pertaining to the methodology of collection and division anticipates the structural content of an Abelian group.

Collection and Division—and Grouping: A Metaphysical Intersection

Gathering particular entities for investigative purposes within the confines of a single circumscribed area is procedurally common to collection in Plato as well as to the formation of an Abelian group in mathematics. But if our present concerns are predominantly philosophical, then we may inquire into the substantive differences between the informal grouping Socrates criticizes in the *Phaedo* and the formal grouping which, according to theoretical pronouncements in the *Phaedrus* and several other later dialogues, represents a necessary stage in identifying Forms as primary causal and definitional agents.

For the later Plato, collection is complemented by division; their interplay produces results—the identification of Forms established by due derivation from gathered particulars. For the mathematician, an Abelian group does work incorporating theoretical possibilities animating the structural intricacy of Abelian groups as elements in a wide variety of modes of inquiry, especially with regard to results expressible at a higher level of mathematical abstraction. For the philosopher, the method of collection and division deploys elements into a group as a prerequisite so that the investigator will be in a position to accomplish a metaphysical end. These programmatic affinities invite juxtaposition.

When we examine the respective structures of (a) the organizational principles assigned to collection and division in Plato with respect to (b) the conditions formally necessary to constitute an Abelian group, the results generate a potent interpretive and critical schematic. From a philosophical perspective, such juxtaposition produces theoretical perspectives pertaining to key factors in Plato's metaphysical ventures embodied in various formulations of collection and division. These comparative implications will, furthermore, reinforce the active, concerted presence of mathematical dimensions—as explicitly codified in the formation of Abelian groups—underlying descriptions of collection and division as a method for producing metaphysical results. In fact, the anticipations of metaphysical and scientific relevance that Whitehead perceived regarding Plato's deployment of basic elements of geometry are mirrored and indeed extended when we shift the mathematical contexts in Plato from geometry to arithmetic and examine these contexts from the perspectives of abstract algebra.

As noted, Platonic interest in collection and division extended from the *Phaedrus* to a series of subsequent dialogues. The statements in these works regarding collection and division render a more detailed account of the structure of that method; we shall examine especially evocative examples of these passages (and repeat the passage from the *Phaedrus* already discussed in Chapter 3) in conjunction with a sample set of elements illustrating the structure of an Abelian group. If the subject matter of collection is limited to numbers and bound variables, then (a) descriptions of the Platonic positions on collection and division juxtaposed with (b) conditions collectively defining an Abelian group will, as asserted above, exhibit affinities underlying the two procedures. We now outline a series of these core affinities.

The passages to be analyzed:

1. *Phaedrus* 265d: "we bring a dispersed plurality under a single form [ἰδέαν], seeing it all together—the purpose being to define so-and-so, and thus to make plain whatever may be chosen as the topic of exposition."
2. *Statesman* 285b:[2] "when a person at first sees only the unity or common quality of many things, one must not give up until [the investigator] sees all the differences in them, so far as they exist in classes; and conversely, when all sorts of dissimilarities are seen in a large number of objects [the investigator] must find it impossible to be discouraged or to stop until [the process of inquiry] has gathered into one circle of similarity all the things which are related to each other and has included them in some sort of class on the basis of their essential nature."

The following account, subdivided into "Collection" and "Division," expands the various systemic characterizations of the elements of Abelian groups to show the extent to which each of them embodies abstract, or as one might put it, metaphysical dimensions. These factors are connected to other, recognizably mathematical elements; as a result, the tendency will be to appreciate their structural characteristics solely in terms of mathematical properties and vocabulary. However, a fairly straightforward expansion of the semantic range of each element of an Abelian group—especially when these articulated elements inhabit a natural language—just as immediately establishes a full range of metaphysical factors and theoretical consequences. We will argue that the principles and concepts that apply to numbers construed in a group context will also apply to metaphysical situations of considerable variety and referential complexity.

Collection

The *Phaedrus* stipulates that collection begins by gathering a "dispersed plurality," that is, particular things available for reflective analysis. The purpose of such gathering is to examine this plurality in its entirety, regardless of how many particulars may appear initially relevant for the investigator to see them all together, under a single Form.

The *Statesman* specifies that such epistemic "seeing" will recognize a unity that gathers "the common quality of many things." This common quality permeating all the gathered particulars will be the Form (also identified in the *Phaedrus* passage as the goal of inquiry). This Form serves as the metaphysical foundation for an account, a definition articulating that Form insofar as the Form has caused or essentially contributed to the existence and determinable structure of those particulars.

Since, however, the items grouped all differ from one another in some respects, the investigator cannot rest content with recognizing the initial uniformity derived from the set of items. Investigation must gather *all* the differences marking the set of items insofar as these differences can also be gathered and identified in terms of a class structure. (This approach suggests that some differences emerging from the set of items will not be so accountable.) The second phase of application begins, therefore, not with the recognition of a kind of immediate sameness marking all the items in the set, but rather with the fact that the items display a variety of differences. The challenge then is to persevere with the fact of these differences so that when juxtaposed with one another, they will display a level of "similarity" that can be attributed to the causal activity of some kind of class.

The reference in the *Phaedrus* passage to "all together" implies that although the set is comprised of particulars, each existing in its own right and independent of any of the other particulars in the group, if each particular is then "added" to any other particular, the result of this operation will establish another particular eligible to be selected for membership in the set for the same ensemble of reasons. The purpose of the grouping is to conjoin each particular with its neighbor (in shared similarity, if not in space and/or time) so that each particular is seen in conjunction with all other particulars. This collective conjoining of individual elements will close the grouping (as required by the conditions for an Abelian group) and also establishes a fundamental sameness between and among all elements in the set grouped (an essential precondition for the Platonic process of collection, ultimately leading to division according to Forms). Closure in this sense guarantees the possibility of at least one or several dimensions of shared qualities determinable between and among all elements in a group. Such qualities, however, do not render any of the elements of the group to be identical to one another.

This type of conjoining presupposes, at an especially fundamental level of Platonic metaphysics, the capacity of one particular to act or to be acted upon by another particular (see *Theaetetus* 248b). For if a plurality of individual entities existed in a completely isolated metaphysical vacuum, then the possibility of addition—as a type of acting and being acted upon involving two (or more) entities—would be obviated, and any attempt to group individual entities for comparative analytical purposes becomes idle, indeed structurally impossible.

If the elements of a class are particulars, then the necessity that particulars fund the collection phase is in contrast to the status of particulars in, for example, the earlier treatment in the *Republic*, where they are described as a combination of being and non-being (Book 5, 478d5-9). The methodology of collection and division holds that particulars have become elements essential as grounding phenomena for the determination of Forms. Indeed, the metaphysician may look to the mathematician for direction in this regard, since for the latter, each member of an Abelian group is fully "real" insofar as each and every member constitutes collectively the complete content of a given group. It may be observed that the range of reality attributed to particulars in the *Republic*, that is, a medley of being and non-being, can nonetheless still be maintained in the context of the methodology of collection and division, but now there is a decided shift in metaphysical emphasis, as it were. From the standpoint of the overall metaphysical system in the *Republic,* particulars are located toward the bottom of the scale of reality; by contrast, when the metaphysician practices collection and division, particulars become the primary exponents of reality in the sense that they provide the "stuff" on which a method can be applied for determining what in those particulars exists and functions at a level occupied by the Forms. In sum, particulars enjoy much more metaphysical respect if viewed from the perspective of collection and division; nevertheless, when all things are considered from a comprehensive and hierarchical standpoint, it could still be maintained that compared to what exists which is "higher" than particulars—the Forms—particulars' hold on reality remains at a level which could be characterized as embodying "not-being" (if "not-being" is allowed to describe significant dimensions of transient, mutable particulars when they have been juxtaposed with eternal, unchanging Forms).

If the multiplicities under scrutiny are limited to countable groups, then no specific number of particulars must be gathered in order to constitute a properly collected set or group. However many elements there may be in a group, this multiplicity is denumerable; that is, it will be at one level subject to being counted. Furthermore, as soon as the set of elements has been determined, any additional items of different membership are excluded, a restriction which holds for the elements in a Platonic collection as well as for the constitution of a given Abelian group. Regardless of the number of items gathered in this uniform way, the collection phase presupposes that these elements, taken as a closed ensemble, will be sufficient to determine a Form.

The necessity of plurality is methodologically pivotal. Thus, gathering a number of distinct instances of certain kinds of realities will yield the constitutive, defining formal structure of all those realities more reliably than would, say, an investigation of any one of those realities taken as a paradigmatic instance of the Form under scrutiny. In fact, the *Statesman* passage emphasizes the need to investigate "many things" [τῶν πολλῶν] to review "a large number of objects" in order to have amassed a reliably numerous and diverse gathering of individual entities as a prerequisite for securing justified metaphysical results identifiable at the level of Forms.

The classes that underlie the process of collection and division are intended to provide parameters for gathering disparate particulars (but related through similarities) throughout all sectors of the natural world as well as whatever human beings have fashioned from this world. However, regardless of the content of a given grouping, the vehicle of collection—here identified as a class—exists and functions in the same way as class functions when the content is restricted to natural numbers. In other words, the type of content of a class does not affect the metaphysical character of the class *qua* class. The implication, which holds for both Platonic methodology and the structure of an Abelian group, is that the gathering instrument—whether class or group—exists in its own right at least to the extent that it retains independence from broad variations derivable from the kind of content subjected to the process of gathering.

The individual elements in a given group may vary in character depending on the nature of the group or groups in which those elements are members. Thus, a given individual element will remain the same in terms of its basic constitution when that element appears in different groups, but its activity within a specifically defined group—a dimension of its reality as that element—will vary depending on the structure of that group. As we have seen, the number 1 functions differently if it is in a set of just two elements compared to its functionality within a group the content of which is a much more numerous plurality. However, once the group is explicitly configured as a group and thereby becomes a fixed, stable mathematical entity, the *relation* between explicitly indicated contours of a group and a given member of that group will also be fixed and invariable to the extent that the *group itself*, existing as *one* determinable mathematical entity, is fixed and stable.

An anticipatory note: if the similarities procedurally present as a precondition for the gathering of particulars refer to particulars all of which are characterized by at least some identical properties, then the natural numbers qualify as the subject matter of collection as formally defined. All items so gathered are numbers, displaying a shared commonality; although numbers may differ from one another in terms of referential quantity as well as relationally, that is, being even or odd, greater or less. With respect to the potential of numbers to lead to the truth (recall the motto of this book taken from the *Republic*), the dialogues emphasize the Forms of numbers taken individually, number by number. However, reflection on the resident structure of Abelian groups shows that numbers considered not only as distinct individual entities but also as collected and grouped into a single closed set can exhibit implications with important metaphysical consequences.

Division

When properly executed, division will produce Forms as constituent factors common to the totality of items comprising a given collection. And when juxtaposed with conditions defining an Abelian group, structural samenesses characterizing metaphysical and mathematical configurations will contribute, both heuristically and substantively, to the establishment of these realities.

According to the *Statesman* passage, all differences characterizing the set of elements must be identified to the extent that these differences can be localized under determinate classes. Once these differences have been stabilized in the sense that their content can be derived from origins based on classes, then it will be possible to focus on the dimensions of reality common to this set of elements and then to articulate how they manifest a "circle of similarity" and, in this respect, how they can be included in "some sort of class [γένους τινὸς] on the basis of their essential nature." This statement of methodology presupposes the active presence of *differences* among the elements of the gathered set; the investigator must review all elements of the group, eliminate all extraneous differences, and concentrate on whatever features remain the same for all gathered members. The result of this complex investigation will constitute the source material ultimately yielding the essence of the gathered elements, an essence expressed through the existence of Forms as foundational and causative agents.

Association. Structural factors defining an Abelian group are both heuristically and substantively relevant to the process of division as it moves toward the identification of Forms and their causal and definitional determination of particulars.

Association establishes a type of limiting interaction between (and among) selected members of the original set of elements in an Abelian group. Thus, a + (b + c) = (a + b) + c provides a procedural parameter, a more *inclusive* grouping within the original and more *extensive* grouping, thereby reducing the number of particulars in play—quantitatively and qualitatively advantageous results with respect to formal inquiry. This kind of systemically available interior grouping places elements in closer contact with one another, thereby facilitating the recognition of sectors of sameness as well as, importantly, differences among the associated elements of the set. If "b" and "c" are closely juxtaposed through association, then the ways in which they are the same and different from "a" will be more readily apparent; the same considerations will hold for "a" and "b" with respect to "c."

Since association can range over all members of an Abelian group, a percipient examination of groupings comprised of limited membership via association will enable the investigating metaphysician to discover more readily the relevant structural aspects of the elements constituting the group. The greater the number of elements in a given group, the greater the increased degree of utility in having a procedural tactic allowing the metaphysician to concentrate on a more limited subset of those elements for the purposes of determining their relevant common features.

Identity element/Inverse element. The identity element and the inverse element—components of conditions defining an Abelian group—can be seen to intersect within the framework of collection and division (in the Platonic sense) as contributors to determining the formal structure of a given group of particulars. More specifically, their joint function emerges as illustrating a *heuristic process* combining elements originally located in the group in such a way as to assist in the process of winnowing out those items not relevant for the identification of defining Forms. This phase of the method is crucial for establishing *only* the Forms present in and defining the reality of the individual particulars collected for investigation.

The *Phaedrus* passage emphasizes the requirement of discerning a common characteristic that pertains to the group of elements, both individually and collectively. The *Statesman* passage provides important additional details concerning how this process of uniform apprehension can be achieved. According to the *Statesman*, the quest to discern the underlying unity common to the group of elements must continue until all the differences characterizing this unity are identified. Furthermore, if the initial apprehensions of the collected particulars reveal only "dissimilarities" present to a large number of these particulars, the investigator seeking the presence of Forms must not be discouraged by the apparent absence of common structures. Rather, due perseverance in the pursuit of defining metaphysical ends will compel the investigator to continue the search ultimately to identify the relevant Forms until "one circle of similarity" has been discerned. This unity by way of similarity, as it were, displays the fact that all elements so gathered collectively produce "some sort of class," with the content of this quasi-class determined on the basis of a sustained awareness apprehending what all these particulars share as their essential nature. However, to be in a position to identify this defining Form presupposes a rigorous investigation of all gathered particulars in order to approximate how this group has been characterized by shared differences. From this methodological perspective, difference and identity go hand in hand.

The *Statesman* passage introduces the factor of dissimilarity within the complex process of discerning Forms common to the gathering of particular entities. We note that dissimilarity, however defined, presupposes multiplicity in that its status as a discernible property emerges when two (or more) entities are juxtaposed with one another so that the properties of one entity will be recognized to differ, at least in one respect (or, perhaps, many respects), from the properties of the other entity (or other entities). Furthermore, as the number of entities in the grouped class increases, it would be reasonable to assume that the number of differences will also increase.

This process of comparison for purposes of discerning similarities and dissimilarities is therefore difficult, if not methodologically treacherous, especially if such a studied review has been undertaken for purposes of ultimately identifying Forms as the essential components of particulars, that is, the constituting samenesses common to all the particulars gathered for this instance of applied collection and division. Plato alerts us in several dialogues to this complicating factor. According to the *Protagoras* (331d),[3] "After all, everything resembles everything else up to a point. There is a sense in which white resembles black, and hard soft and so on with all other things that present the most contrary appearances." And the *Sophist* (231a-b)[4] delivers sage advice in this area of deep-seated metaphysical investigation: "a cautious person should above all be on guard against resemblances; they are a very slippery sort of thing." Since the *Statesman* stipulates that the investigator must continue the process until "all the differences" characterizing the total number of elements collected have been identified, the level of concentration and close examination of the elements under scrutiny must be as methodologically precise and analytically accurate as possible.

This phase of the process, a careful exercise in culling all non-relevant elements from the group, points to the crucial *heuristic utility* of the inverse element and identity element as deployed in a given instance of an Abelian group. These related elements come into play on those occasions when examining the consequences of a certain kind of interaction between or among elements of a group results in the elimination of one of these elements—or, concisely put, "undoing the operation." In these cases, the effect is the "undoing" of the process of collection with respect to non-relevant elements (given that the ultimate purpose of the collection was to produce a specific result, common to all members of the group but existing at a higher level of generality and identified as such by division). The consequences of the inverse element of a group, if viewed from a parallel metaphysical perspective, result in eliminating that item from membership in a collection. The inverse relation itself remains an essential feature in a group. However, the presence of inversion, when heuristically applied to a certain item, effectively removes that entity from consideration as a relevant member of a collection in the Platonic sense.

This procedural undoing will occur, if and when necessary, only after the operation (i.e., addition) has juxtaposed elements of a group as collected so that investigation will reveal that a feature or features of one or more elements are not conformable to the general structure of the remaining elements in the group. The presence of this feature of an element requires that the element be "subtracted" from the totality of elements in the original group. If an element (or elements) does in fact possess and exhibit this difference, then, as just noted the element as such, as a whole, is disqualified from membership in the group. The assumption is that a foundational Form must be common to *all* relevant aspects of *all* members of the group; if a projected Form is in fact incompatible with some aspect of the individual constitution of a particular (or particulars), then the particulars so affected must be removed from methodological consideration.

The relevance of 0 as an integral element in the applied methodology of collection and division becomes apparent in this setting. It would be as if adding one member characterized as possessing a certain property to another member *not* containing that difference results in zero (a result that produces the cancelling in some but not all Abelian groups); that is, a kind of absent specification and relevance as far as the purpose of the original process of collection is concerned. When such cases arise during the investigation of the members of a given group, we should underline the fact that the existence of both individuals as such is unaffected; the excluded particular continues to exist as that particular however it may be metaphysically constituted in its own right, but its existence is no longer relevant to the specification of foundational Forms of that specific instance of collection and division.

The related Abelian factors of the identity element and inverse element therefore effectively summarize the "winnowing" phase of Platonic methodology by contributing a crucial heuristic perspective on the application of collection and division. It may (and doubtless will) happen that an element initially appears to belong to the aggregation of particulars grouped by virtue of apparently common properties; however, upon careful and critical scrutiny, we may determine that this element exhibits non-relevant characteristics (certain specifiable differences compared to the other gathered elements) such that its membership in the group would no longer be viable with respect to the goals of inquiry.

How is such critical scrutiny employed in order to achieve this end? We recall that the inverse element in an Abelian group in effect undoes the operation. Now transfer the result of this process to a setting defined according to Platonic principles. If two elements of a collection are juxtaposed in such a way that this methodological move "undoes" the operation of addition, then the juxtaposition has revealed that the two elements, if considered from a strict metaphysical standpoint, are different from one another in the requisite formal sense, that is, with respect to subsequent theoretically possible division for identifying essential Forms. Realization of this incompatibility becomes apparent as a result of taking the inverse element and identity element—necessary features of an Abelian group—and applying them, in heuristic fashion, to specific elements in conjunction with the collection phase of Platonic metaphysics.

Any two elements in the collected set as individual totalities will differ from one another in at least one respect (this is a metaphysical given). These differences must be duly noted in order to situate the elements so that what is common or shared between them can be identified, then properly designated as derived from or caused by a Form. But if the differences marking the juxtaposed elements happen to be multiple, complex, and subtle (as the *Protagoras* and *Sophist* have warned!), then a complete recognition of these differences may be difficult to achieve. In the midst of such metaphysical uncertainty, the inverse element and identity element provide a groundwork formalization procedure for examining elements according to modes of opposition or negation. This procedure effectively functions as a kind of shuffling or variation for purposes of determining, in a precise way, how any two gathered elements are the same as and different from one another.

The elements in an Abelian group remain (since the inverse element and identity element represent defining features of such grouping) and so do the elements of an exercise in collecting from a Platonic perspective (insofar as these elements continue to enjoy their self-sustaining character as particulars). This element and perhaps both elements, when first collected, appeared to be potential candidates for inclusion in the final, gathered set. However, upon due examination, the differences characterizing the original element (or elements) emerge as recognizably inappropriate for inclusion in this exercise in collection. This integral result, a necessary feature of the methodology of collection and division when successfully employed, has been produced by an investigative process based on the modes of active negation and comparative variance derived from heuristically adapting the same kind of inverse features proper to the basic structure of an Abelian group.

As noted, the now-excluded elements will continue to exist as such, in their full reality, but they have become residents in, so to speak, another world, at least from the standpoint of the "world" (or set) generated by the process of collection intended initially to provide a base of particularity for establishing the identity of Forms defining the members of that world. Thus, any excluded particular may become relevantly gathered *in another collection*, which may be the source for identifying *different* Forms, but not in the specific metaphysical setting characterized by the set of all particulars coexisting *in this collection* and (after due process) allowing the identification of Forms available only from these particulars.

The Recognition of "Higher" Realities

The comparative analysis in this section has identified counterpart affinities and functions in the juxtaposition of (a) requirements for an Abelian group and (b) structural elements concerning collection and division as drawn from passages in later Platonic dialogues. Now consider their respective purposes, summarily stated—for collection and division, to produce Forms; for Abelian groups, to function as a foundation for the development of more abstract and complex structures, thereby contributing to the ongoing advance of mathematical wisdom in matters algebraic.

In the *Phaedo*, Socrates insisted that investigation into the causality of numbers must take the good into account (99c); however, he provided scant detail on what is available to human cognition as far as the good itself is concerned. Socrates does say in this dialogue that the good "holds things together." To illustrate this function, the introductory discussion of the good in Chapter 1 pointed to contexts in the account of numbers and related mathematical elements which may be taken to illustrate this foundational type of interrelatedness and subsequent bonding. Furthermore, given the testimony of the *Republic* concerning the good, we may assert that this high-level connectivity will permeate all reality. If the good also functions at the same level within the theoretical confines of the various configurations of collection and division—as Socrates explicitly says that it does for the philosophical ends described in the *Phaedo* and *Republic*—then all the processes introduced in subsequent dialogues (and illustrated by the passages from those dialogues quoted above) will receive their ultimate measure of theoretical "rightness" from the good.

For both the Platonic approach to metaphysics as well as the mathematical systems embodied in Abelian groups, the direction of the goal of inquiry is identical—to "ascend" to higher levels of reality, where "reality" refers to what exists on more abstract planes of significance and theoretical structure. Indeed, if the phrase is allowed, we may identify the speculative development as "shared directionality"—theoretical motion from a lower to a more elevated level of abstraction. Ascendancy for Plato culminates in a single reality, the good, beyond which nothing exists or can exist, whether from the standpoint of metaphysics or any mode of normative inquiry. By contrast, mathematical developments based on Abelian groups initiate formulation as well as articulation of ever-increasing levels of mathematical abstraction and ascending algebraic complexity—fields, rings, and vector spaces, all types of group theory, leading to topological groups and Lie groups.

In this regard, the arena of abstract algebra mirrors the ascent toward the good in the Platonic metaphysical scheme. However, whether a similar upper limit to the theoretical possibilities inherent in Abelian groups and group theory in general exists would not be possible to determine *a priori*, as it were. It seems safe to assume that the mathematical imagination will function without predictable limitations in this area of algebraic development. It also seems safe to conclude that the results of mathematical inquiry and demonstration based on Abelian groups will, in parallel fashion, be "held together" by the structural solidity of its original defining axioms—the counterpart of the good from a mathematical perspective. This foundational safeguard will be in place regardless of the subsequent development of proofs, especially as these more complex, increasingly abstract, and rarified (i.e., less intuitively accessible) theoretical levels have been identified and their individual structures described. As history tells us, mathematicians will do what they will do, quite apart from the broader theoretical and speculative settings that students of philosophy in general and metaphysics in particular may erect to house the efforts of practicing and groundbreaking mathematicians from the standpoint of interests in producing some sort of fully comprehensive system.

Mathematics and Metaphysical Vision: A Look Ahead

The appeal to the good as a functioning component in mathematical contexts may seem out of place and, in terms of philosophical respectability, quixotic at best. Perhaps. The reader is asked to hold off judgment in this regard until the development of the arguments laid out in Part II. And to initiate a transition to Part II, we return briefly to Whitehead for a final thought on the major strategic move in Part I—juxtaposing Platonic approaches to numbers as bearers of metaphysical import with the modern development of Abelian groups in mathematics.

When Whitehead said in *Process and Reality* that "Plato would have expected" the transformation of mathematical elements (i.e., certain types of triangles) into a very different theoretical medium requisite for the concerns of modern science, he was, of course, not claiming to have privileged insight into the hypothetical thought processes of Plato as a human being. Rather, Whitehead was interpolating the experiential likelihood that certain interpretive conclusions based on principles enunciated in the *Timaeus* would have been positions Plato could reasonably have foreseen and endorsed. Whitehead based this theoretical scaffolding on subsequent scientific inquiry, studying activity more than two millennia distant from the Platonic worldview embodied in the *Timaeus*. In this book, we have so far pursued a similar exercise in high-level comparison; the difference is that the context of comparison is defined by the juxtaposition of arithmetic and philosophy, metaphysics in particular, rather than by geometry and physics, as it was for Whitehead. But although considerable (indeed, even greater) historical separation remains between the subject matter of the mathematical and philosophical comparisons drawn here, the parallels between the two disciplines are consonant.

The narrative strategy of this book has shown that aspects relevant to the formation of Abelian groups were presaged in the *Phaedo*'s account of the generation of numbers and inchoate in the formulation of the Platonic methodology of collection and division. These aspects revolve around connections between particular phenomena and the Forms, especially (a) the identification and "gathering" of individuals that share properties for purposes of duly informed investigation eliciting common and more abstract structures as well as (b) the processes involved in subtracting individual elements which, in the last analysis of the group of items under scrutiny, did not in fact provide relevant source material. Consider this connected ensemble of elements, but for now, solely from a mathematical, and in particular algebraic standpoint. These factors collectively represent necessary features of an Abelian group. As a result (and with Whitehead's derivation standing as an exemplar), we may wonder about Plato's possible reaction to the following intricate interplay of circumstances: first, mathematicians in subsequent eras concentrated on purely mathematical implications from characteristics and approaches to that phase of the *Phaedo*'s metaphysics concerned with the generation of numbers; second, they then fashioned a schematic structure—the Abelian group—preserving fixed elements in a theoretical account of numbers but with a flexibility that allowed considerable development in understanding the groupings of a variety of algebraic realities.

Now imagine, finally, Plato observing the array of complexity and breadth inherent in the structure of an Abelian group and also appreciating the affinities (sketched above) consonant with his own abstract development of collection and division. We may then say, in a manner more restrained than Whitehead's turn of phrase but sharing in the same kind of positivity, that we can also imagine Plato not being at all surprised that mathematics, at a later stage of its historical development, would have established the formal conditions for Abelian groups and their resulting fecundity in analyzing algebraic structures at an increasingly abstract level of generality. Indeed, if approached from an accommodatingly broad perspective, the method of collection and division—although complex (especially as diversely represented in later dialogues)—incorporates the core of concerns animating the structural character of an Abelian group. However, given the relatively non-intuitive character of the identity element as well as the inverse element, two requisite features of Abelian groups, the degree of Platonic anticipation of such mathematical innovation would be ratcheted down from "expectation" (in the Whiteheadean framework of description) to "a lack of surprise." This reduction in anticipatory intensity would reflect the immense dimensions of mathematical imagination displayed throughout history, as well as, in particular, the necessarily limited content of what Plato could appreciate about the complex, richly diverse reality of numbers and mathematical entities, especially as concentrated in the determinate structures of Abelian groups and their developments in the subsequent history of mathematics. Plato was a philosopher, not a professional mathematician. It nonetheless seems fair to conclude that his abilities in the former modes of inquiry allowed him to see deeply into the foundations of the latter and its place as an especially profound register of the possibilities inherent in the processes of creative thought proper to the human spirit.

Anticipation and a lack of surprise are perhaps intriguingly imagined details concerning Plato's basic humanity regarding the interplay between mathematics and philosophical endeavors. It would be prudent to recall, however, that as far as the fixed record of the extant texts is concerned, Book VI of the *Republic* locates mathematics in general "below" the Forms (and by extension the good) in Plato's own metaphysical hierarchy. How then is the good relevant to mathematics in any sense, much less with respect to the degree of reality exhibited by the Forms?

The narrative approach in Part I has been almost exclusively expository (two exceptions: the brief, speculative discussions comprising what we labeled the schematic preamble in Chapter 2 and the series of arguments offered in Chapter 3 concerning the complex metaphysical status of the number one); then, with an essential boost from Whitehead, we set forth the positions and resulting intersections involving Platonic metaphysics and elements of a very different metaphysical emphasis on process (Chapters 1 and 2). The subsequent identification of commonalities emerging from the juxtaposition of the method of collection and division with Abelian groups (Chapters 3 and 4) stands as basic components which, with development, will have important ramifications for the relevance of mathematical realities regarding the critical assessment of metaphysical theorizing. It is at this juncture where the Platonic vision of the good assumes a dominant and directive role—its comprehensiveness as a fundamental principle encompassing inquiries into *all* sectors of reality as well as that more limited mode of reality enveloped by mathematics, marked as mathematics has been throughout history by ongoing and highly diverse theorizing and demonstration.

To assert, develop, and justify approaches to mathematics and its potential role in metaphysical analysis, all from the standpoint of theoretical applications derived from the good, stands as an ambitious project of considerable scope. A project of such dimension requires a broad, well-articulated foundation to support additional perspectives on such a theoretical venture. It is crucial to note, then, that although we can initially ground such speculation in a Platonic approach to the good, it will become strategically necessary to shift and refocus its argumentative direction.

Part II pursues this objective. We will initiate an account of this more expansive setting by following an indirect route via one of Plato's most pointed critics—his student Aristotle. Arguing from a related but in many ways very different metaphysical perspective, Aristotle introduces lines of criticism the implications of which will extend the range of influence established by the intersection of mathematics and metaphysics sketched so far in Part I. As such, this Aristotelian prelude (Chapters 5 and 6) becomes an integral phase in the process of articulating this intersection. We may also emphasize now, at the conclusion of Part I, that throughout this section of the book, the argument never loses sight of the good and what its reorganization and implementation contribute to the direction and details of the final positions developed in Chapters 7 and 8, the book's speculative conclusion.

Notes

1 For a more rigorous exposition of the basic elements of Abelian groups, see Robert J. Valenza, *Linear Algebra: An Introduction to Abstract Mathematics* (New York: Springer, 1993), 18–19.

2 Skemp's translation is slightly modified.

3 Translated by W. K. C. Guthrie, *Protagoras*, in *The Collected Dialogues of Plato*, ed. Edith Hamilton and Huntington Cairns (New York: Pantheon Books,1961).

4 Translated by F. M. Cornford, *Sophist*, in *The Collected Dialogues of Plato*, ed. Edith Hamilton and Huntington Cairns (New York: Pantheon Books,1961). Slightly modified.

5

Aristotle and the Metaphysics of Numbers

Overview—Part II

In Part I, we argued that the Platonic approach to the metaphysics of number in the *Phaedo*, if taken in conjunction with the *Phaedrus* and *Statesman*, embodies a theoretical affinity with Abelian groups, an important area of investigation in modern group theory in mathematics. We then showed that aspects of the defined structure of Abelian groups can be correlated with elements characterizing the later Platonic interest in identifying Forms by means of the method of collection and division. Regardless of the historical span of time separating their promulgation, uniformity of structure correlates these mathematical and metaphysical ventures. As a result, we have secured a foundation rooted in mathematical considerations for exploring and exploiting this confluence of shared elements for purposes of metaphysical analysis, that is, critical evaluation and theoretical expansion of positions advancing philosophical conclusions addressing metaphysical ends.

As a prerequisite for pursuing an inquiry of this sort, we restate for the sake of emphasis the following as a principle of interpretation: a mathematical environment of modern origin contains notions incorporating foundational perspectives suitable, when stated in the abstract, for analyzing a methodology formulated in antiquity, aimed at articulating purely metaphysical ends. Part II develops a series of implications of this principle by reviewing and analyzing texts from Plato, Aristotle, and Whitehead; the discussions in Part II pursue this exercise in critical and "interdisciplinary" reflection in ways that will establish positive interpretive strategies. Broadly stated, the primary burden of Part II is to demonstrate that structural characteristics proper to mathematical inquiry provide various useful points of departure for investigating metaphysical positions. The demonstration of these emergent characteristics will include developing a speculative expansion of the divided line (*Republic*, Book VI) insofar as the Platonic metaphysical vision embodied in that famous image incorporates mathematics within a more inclusive, comprehensive metaphysical hierarchy.

Concerted appeals to Aristotle's keen critical eye will assist in establishing positions underlying the formulation of these generalizations. Indeed, key sections in Aristotle's *Metaphysics* provide source material integral to the requisite foundation for this kind of inquiry. During the initial phase of its development, this theorizing constitutes a series of exercises in comparative metaphysics. So far, Aristotle has remained in the wings (other than his intriguing claim about the importance of beauty and the good in mathematical contexts, one of the book's mottos). Now, however, we will take advantage of the fact that Aristotle thoroughly examined and critically evaluated the Platonic approach to number as presented in the *Phaedo* as well as the *Republic* and a related position adumbrated in the *Philebus*.

Aristotle's criticisms in the *Metaphysics*, informed by many of his cardinal metaphysical principles and notions, raise a variety of seminal issues regarding Platonic conclusions in this area of inquiry. For Aristotle, these conclusions lack cogency as contributions to understanding number and mathematics in general primarily because they were derived from inadequate metaphysical foundations. Furthermore, as we shall show, the effective scope of this critical temperament can be extended to the subsequent Platonic approach to the Forms by means of pursuing the method of collection and division. With the reasoning underlying this critical background in hand (also exploiting inferences drawn from a revamped notion of the good), we then formulate an argument demonstrating that the student of metaphysics may incisively apply this evaluative schematic—derived, we recall, initially from extended criticisms of Platonic positions on the nature of mathematical realities—to a broader range of metaphysical ventures than those appearing just in Plato's dialogues.

As a preamble of sorts, it will be instructive to describe, at a fundamental level, the notional background for the investigations discussed in the opening phases of Part II, Chapters 5 and 6. The following elements are common to (a) collection and division as well as (b) the metaphysical character of numbers—class (associated with unity in a necessary structural sense), a unit entity or particular, plurality as identified by a specific number of unit entities, identity and difference, property, and the implicit presence of relationality (for example, as found numerically in the "greater than" and "lesser than" relationships and also in the sequential character of numbers).

This broad range of commonalities is important to note, since its extensive and varied theoretical elements justify posing the following questions as parameters for developing the main argument of the book and, at this juncture, as a key directional pointer for Part II: are Aristotelian criticisms of the Platonic position on numbers (and their various metaphysical dependencies) also relevant to the Platonic practice of collecting and dividing a denumerable set of particulars (as a methodology for determining the presence of Forms "in" these particulars)? Are there parallels, or at least relevant similarities, between the Platonic view of numbers and the Platonic deployment of elements animating collection and division as a methodology for producing fundamental metaphysical ends? Finally, if such parallels can be derived from these elements and are expressible according to metaphysical specifications, can the Aristotelian critiques of Platonic positions on number, driven as they are ultimately by metaphysical considerations, also be employed as critical wedges in assessing the cogency of the metaphysics underlying the methodology of collection and division? In sum, do criticisms of mathematical positions facilitate parallel criticisms of metaphysical positions?

The answer to these questions taken collectively is, in a word, yes. We will show (in Chapters 5 and 6) that the background affinity wherein mathematical and metaphysical considerations intersect, illustrated so far in a purely Platonic setting limited to the analysis of number, also pertains to Aristotelian criticisms applied to Platonic metaphysical speculation (couched in the methodological details characterizing collection and division as well as the intended abstract objects discernible after successful use of this method). Thus, as a procedural strategy, we will juxtapose (a) various blended mathematical and metaphysical elements in Aristotle's array of criticisms with (b) basic metaphysical dimensions actualizing the conjoint methodology of collection and division.

Aristotle's critique of Platonic philosophy of mathematics appears in adumbrated form early in the *Metaphysics* and in much more concentrated and sustained analyses comprising the *Metaphysics*'s final two books, M and N. These criticisms will, upon analysis, clarify important features of the processes involved in the application of collection and division, and also raise substantive issues regarding the cogency of this methodology. Concisely stated, Aristotelian mathematical criticisms have important metaphysical implications in the immediate context under scrutiny. And as we shall see, these implications typically revolve around a variety of divergent (and often unconsidered) senses of unity, one of the core concepts of this book.[1]

We proceed by selecting various Aristotelian criticisms of Platonic philosophy of mathematics and laying out in some detail their structure from the standpoint of dependency on common metaphysical components (Chapter 5). Then, with this background argumentation in hand, we will redirect those criticisms at related generalized areas in the Platonic account of collection and division (Chapter 6). The fact that Aristotle did not intend the stated mathematical criticisms to challenge metaphysical aspects embedded in the methodology of collection and division does not affect the import of these criticisms as far as establishing a slate of informed philosophical concerns critically relevant to that methodology. In fact, the range of results produced from the inquiries in Chapters 5 and 6 will demonstrate that theoretical aspects of collection and division, when deployed in notional conjunction with elements of Abelian group theory and then subjected to adaptations engendered by Aristotle's critical scrutiny, represent another fertile interplay of metaphysical evaluation inspired by and originating within mathematical considerations.

If we follow the lines of argument raised in Books M and N as discussed below, the contention is that we will achieve insight into procedural as well as metaphysically substantive aspects of collection and division. Whitehead read Plato as a proto-mathematical thinker in his use of mathematical elements, triangles, for metaphysical ends in the *Timaeus*' account of the cosmos fashioned in all respects "for the best." This rarified dimension of Platonic speculation anticipated, in its point of departure, discoveries in mathematics (as well as physics and chemistry) that did not appear for two millennia. In this book, we are in a sense shifting the direction of inquiry. The following analysis originates with predominantly mathematical concerns (raised by Aristotle) but culminates in a more nuanced, complex understanding of a methodology for achieving metaphysical ends (developed in later Platonic dialogues).

Chapter 4 showed how this methodology is developed by the use of interlocking elements which parallel the structure of group theory as pursued in modern mathematics. Aristotle's extensive critical analysis—with related but provisional responses to those criticisms offered as commentary—reaches a level of detail and depth which can fairly be described as metaphysically foundational. We may take these exercises, in addition to their inherent interest with regard to Aristotle's view of what Plato intended to accomplish by collection and division, as an introduction to the even more inclusive accounts developed in Chapters 7 and 8. The book's concluding chapters demonstrate how an articulated and in some ways considerably altered approach to the Platonic notion of the good can accommodate and orient a vision of mathematics integrally related to the "upper" echelons of metaphysical speculation. In this regard, it will be of special interest to examine the text in *Metaphysics* Book N where Aristotle himself explicitly mentions beauty and the good in a mathematical environment. Implications from this passage, in conjunction with the review of Aristotle's extensive criticisms levied against Platonic positions, will contribute to more explicit and feasible ways to approach the function of the good in various areas of metaphysical inquiry when we view that function from the perspective of mathematical investigation. In fact, the investigations pursued in Chapters 7 and 8 will significantly broaden the relevance and practical application of elements in mathematical positions with respect to the critical and creative examination of a broad range of metaphysical ventures.

Introduction—Chapter 5

The Aristotelian criticisms discussed in this chapter are aimed at the Platonic position on the generation and mode of existence of numbers. As indicated in the Overview, the positions standing as the target of Aristotle's critique appear primarily in the classic middle period dialogues, the *Phaedo* and *Republic* (and, indirectly, the later *Philebus*). These dialogues advance positions on numbers that depend ultimately on various metaphysical notions and principles, for example, the Forms as canonically identified—a line of thought that Aristotle rejects with a veritable barrage of criticisms. There are, however, additional and metaphysically important arguments against Platonically oriented positions on mathematical matters (positions that are textually complex in ways to be discussed). All these criticisms incorporate a variety of notions common to many traditional patterns of metaphysical inquiry.

Aristotle's critical strategy includes much more than targeting problems aimed at the classic issue of how it is possible for particulars to "participate in" Forms. A prerequisite for appreciating the direction and detail of Aristotle's criticism is, therefore, a summary of the theory of numbers which represents systemically what Aristotle apparently believed Plato, or at least the representatives of Plato in the Academy, accepted as their "doctrinal" position. The textual status of this position (with respect to Aristotelian criticism) will be discussed after we present its basic components.

Numbers, Units, and Unity in Plato

Texts in the *Phaedo* indicate that Plato's initial understanding of numbers was that their existence depended on participating in appropriate Forms—the number 2 exists as a number because it participates in the Form of duality. However, Aristotle's criticisms of Platonic theories of number are often aimed at a different position, the basic elements of which we will outline in this section.

The relevant realities (typically identified as "Form numbers" or "ideal numbers" in the secondary literature) contain "units" [μονάδι]. These units are "combinable" [σύμβλητοι] in certain specific ways. Julia Annas' statement is clear and useful: "Units are combinable in a number if they can be 'combined' or grouped to make up that number and no other. The units in 2, for example, are differentiated by the fact that they cannot be combined together to make up any other number" (JA, 17–8). W. D. Ross expands on the "unit" dimension of numbers:

> The ideal numbers are simply the natural numbers, i.e. the universals twoness, threeness, &c., of which all groups with two, three, &c., members are the particular instances. From their nature as Ideas it follows that they are specifically distinct and incomparable, i.e. incapable of being stated as fractions one of another. Twoness is not the half of fourness. Nor is a natural number an aggregate of units. If, therefore, the Platonists had been true to their principles, the question which Aristotle presses on them, whether the units in ideal numbers are comparable [i.e., combinable, following Annas' translation] or not, would have fallen to the ground. Their answer would have been that there are no units in ideal numbers. It is *possible* [italics in text] that Plato held this view, but certainly some of the Platonists did not. Aristotle says (1080^{b}8) that the views (*a*) that all units are comparable, and (*b*) that the units in each number are comparable to each other but not to those in any other number, both found support among the Platonists, but he does not expressly assign either to Plato.[2]

This extended passage provides background on the structure of units in the context of numbers and also, importantly, indicates that the position under scrutiny may not have been advanced by Plato himself, even if it had been the position of some Platonists in the Academy. The primary difference between the *Phaedo* position on numbers and the subsequent position on "Form-numbers" is noteworthy. In the *Phaedo*, the Forms standing as the origin of numbers are themselves uniform, as are all Forms according to that dialogue; in the later position, by contrast, each of the Form-numbers is internally differentiated into units which presumably account for the distinctive numerosity of a given number (the Form-number "two" has two units; the Form-number "three" has three units, etc.). Is it possible to clarify how we are to understand this appeal to "units" with respect to their function in determining the unique character of each number? Ross offers a response of sorts: "*Monadikos arithmos* thus clearly means 'number made up of abstract units,' not just units" (Ross, Vol I, 15, fn20). We will discuss below a possible range of meanings of "abstract" in this context, since it will become evident that Aristotle's criticisms often turn on the kind of unity—"abstract" if Ross is correct—that can be ascribed to each of the units under scrutiny.

The point to be noted at this juncture is textual and historical—the direction and detail of Aristotle's critical responses to the notion of Form—numbers are aimed at positions that were alive when Aristotle was actively dealing with these issues, but which also may not have been advanced by Plato himself. Annas neatly summarizes (JA, 163) the Aristotelian strategy of this section of Book M: "The main part consists of a refutation (1080^{b}3–1083^{a}17) of the Platonic conception of unique Form numbers each of which is a collection of units of a special type." This extended exercise in refutation is pursued even though passages from the extant dialogues are unavailable to support the straightforward assertion that Form-numbers exist and function as Aristotle believes them to function. As a result, our concern is with Aristotle's treatment of seemingly viable arguments concerning the philosophy of mathematics, but not necessarily with canonical arguments of Plato. We may wonder whether granting serious critical attention to such positions, given their seemingly shaky textual bases, is warranted; Julia Annas has pointed out relevant comments on this somewhat convoluted authorial situation:

> To hold that Plato did not have such an arithmetical conception of number, one must also hold that all Aristotle's criticisms are either massively misguided or completely unfair, gratuitously presenting Plato with a position totally unlike any of his actual views, merely in order to refute it. This has been maintained, but any plausibility it might have can hardly survive a serious reading of the arguments in [Books] M-N. They are detailed, technical, patient, and incisive; it is not likely that the person capable of producing them either stupidly aimed them at the wrong theory, or indulged in the pointless (and easily detectable) ruse of falsely attributing to Plato the position they did refute. (JA, 16)

The vigorous tone suggests that Annas leans toward the view that Plato himself, not just scattered Platonists, maintained in some manner the views Aristotle considers and criticizes. But, again, her description of Aristotle's arguments—"detailed, technical, patient, and incisive"—is, for our purposes, the key aspect of this passage. We will watch Aristotle confront positions (regardless of authorship and allegiance) that he believes are incorrect, then learn how, at least by his lights, criticisms weaken if not refute these positions. The arguments themselves as well as Aristotle's critical rejoinders to these arguments collectively provide useful instruction on fundamental issues in the philosophy of mathematics and serve as introductory analysis on connecting the implications of this instruction to even more fundamental issues in metaphysics.

To buttress the theoretical relevance of this approach, consider briefly a more formal aspect as it pertains to the same textual circumstances: basic syllogistic logic teaches that an argument can have false premises and yet yield a true conclusion—such an argument could be valid (with respect to standard logical form) but not sound. As a rule, these arguments are not considered philosophically cogent due to the falsity of the premises. This technical point is mentioned here only to reinforce the relevance of Aristotle's thinking given the project at hand—the metaphysical dimensions of classic philosophy of mathematics and their utility for metaphysical inquiry in general. What we care about in this context is not the accuracy of Aristotle's scholarly acumen in terms of whether or not his premises correctly capture what Plato may have held regarding the metaphysical constitution of numbers, but whether, in the course of the reasoning Aristotle employs in examining these premises, metaphysically relevant aspects of the positions so scrutinized become open to critical investigation.

Procedural Comments

The complex scholarly considerations that apply to the study of Aristotle's texts are well-known. As a prelude to investigating Aristotle's critical assessments of Plato's philosophy of mathematics, it is therefore useful to remind the reader of factors bearing on this kind of analysis. The *Metaphysics* is a *locus classicus* for these difficulties. Aristotle discusses mathematics early in the *Metaphysics*, that is, Book 1, Chapter 9 (esp. 991^{b}9–992^{b}17). His criticisms of Plato derive from Aristotle's understanding of unity (and much of this material reappears later, in Book M). Also worth noting is Ross' point (Vol. II, 495) that Aristotle nowhere gives a positive theory of number (Ross suggests that the nearest such account is in Chapter 3 of Book M), but as already emphasized, our concerns here are with Aristotle's slate of criticisms of the Platonic position and what can be learned from these arguments. Whatever Aristotle's own position on this topic may have been (or what could be inferred as a cogent position in this area from texts indicating what Aristotle did say) is not part of our current range of interests.[3]

Five separate critical themes are introduced in this chapter. Each of the five sections includes a heading and a capital letter indicating that all discussions in a given section are devoted to a single, limited line of criticism (although some sections, e.g., A, contain a cluster of arguments). All arguments examined are included in their entirety as they appear in Books M and N. The sequence of themes introduced here, A through E, generally follows the order of texts as they appear in Books M and N. We should note, however, Julia Annas' editorial comment (JA, 216) on the progression of thought in passages 1092^{a}17-^{b}8: "They do not belong very well either with what precedes or with what follows—another sign that M-N is a collection of notes put together by an editor." If therefore we follow Annas' textual reservations about the entirety of the texts compiled and now read as Books M and N, then any slight variations in (a) the sequence of criticisms as deployed in this chapter when compared to (b) the order of topics in the existing texts of M and N should not cause concern. In that regard, we underline the point (first made in the Preface) that our approaches to the texts of the philosophers considered in this book are philosophical and speculative rather than exclusively governed by scholarly considerations in traditional senses. Finally, comments from either Annas or Ross (or both) will be included as reactions to the cogency of Aristotle's reasoning. Additional reactions of my own will also appear when relevant.

Immediately following the designation of themes A through E, the relevant elements of Abelian groups appear (in brackets) as they have provided substance for the various arguments. Indeed, an examination of the details of Aristotle's reasoning indicates that in several instances, these elements work in harmony with one another. Such intimate systemic integration is only to be expected, given the equal necessity applicable to all conditions in the formal definition of an Abelian group. Additionally, this characteristic of theoretical integrity underlines recognition of parallel critical pursuits when the generalized factors of Abelian groups, initially a purely mathematical phenomenon, are directed to the explicitly proposed operations of collection and division as a putative contribution to metaphysical theory. In fact, deploying the textual details of Aristotle's arguments under headings that fit the structure of Platonic collection and division reinforces the relevance of analyzing and assessing the energy latent in Aristotle's criticisms from the standpoint of generalized metaphysical concerns. Furthermore, this approach sharpens the senses in which Aristotle's critical temper accentuates problematic areas in the strictly mathematical dimensions of Platonic thinking. Illustrating the extensiveness of these mathematical/metaphysical parallels will be the primary burden of Chapter 6.

An additional thought worth keeping in mind by way of the context of inquiry: We noted earlier Julia Annas' thumbnail evocation of the ruthlessness of Aristotle's arguments against Plato in Books M and N. It may be a small but perhaps appreciated gesture of basic humanity to recall what Aristotle said in the *Nicomachean Ethics* (Book I, Chapter 6) just prior to his sustained and, indeed, severe critique of the Platonic idea of the good, that is, that "piety requires us to honor truth above our friends" or, as sometimes rendered, we hold friendship dear but truth dearer, in context a sentiment of decided narrative intensity and atypical of Aristotle's usually clinical, often blunt modes of expression. The matter under present scrutiny—a series of issues in the philosophy of mathematics—might seem initially to be a context of inquiry lacking in long-range human utility and thus existing as a textual phenomenon inherently distant from exciting the kind of existential urgency that causes some academicians doing philosophy to lacerate their opponents in print. Aristotle reminds us that it is integral to the discipline of philosophy that its exponents want to assert truth about all such matters even if to do so adequately requires laying bare the errors of "those we hold dear." It will follow then as a matter of course that friendship remaining subservient to truth is a procedural principle that also applies to competing philosophical views on mathematical realities.

The Status of Classes—Criticism A [Closure/Identity Element]

The *Phaedo* contains what could be labeled the "standard" Platonic position on the metaphysics of numbers, and we have discussed it at various junctures in Part I. Aristotle did not ignore the claims about numbers in the *Phaedo*, although he aims the majority of his criticisms primarily at aspects of the realities designated as Form-numbers, positions at least approximated in later Platonic work. The approaches to the *Phaedo* nonetheless illustrate the strategic lines Aristotle will employ in his criticisms and also bring to the forefront several fundamental dimensions of metaphysical theorizing when its elements enter a mathematical setting. This interdisciplinary strategy justifies the use of "class" in the heading of this section as a more neutral, metaphysically indeterminate name for the target of analysis in this section.

The initial phase of Aristotle on mathematical positions stated in the *Phaedo* is not strictly an argument; however, its directness leaves no room for uncertainty regarding the general attitude driving his critical assessments of those positions. At *Phaedo* 101c, Socrates asserted that the number 2 is what it is because it "participates" in twoness; at *Metaphysics* 1079^{b}25, Aristotle responds with a sweeping rejection of all such theorizing: "And to say that they [Forms] are patterns and the other things share in them is to use empty words and poetical metaphors." This characteristic bluntness encapsulates the Aristotelian reaction toward attempts to account for the existence of particular entities, including numbers, by appealing to a special relation between these entities and certain members in the world of Forms.

This resort to what might be described, with Aristotle, as "metaphysics via metaphor" puts pressure on the appeal to the unity of a Form of duality in the *Phaedo* account of numbers. Aristotle attacks the putative unity of a Form from different theoretical directions; thus, the considerations introduced in the initial criticism contain a medley of arguments, a sort of theme and variations, developing a series of pointed, reasoned criticisms pertaining to the fundamental notion of a class insofar as it functions as a pivotal and necessary element in a metaphysical account of numbers. The arguments pursue three perspectives on Forms: (a) the participation relation; (b) Forms understood with respect to the fundamental metaphysical processes of being and becoming; (c) the extent of the Forms (in two senses—comprehensiveness with respect to a given type of reality, and comprehensiveness with respect to all realities). As a result, this section has been subdivided into phases, each detailing what for Aristotle is a problematic aspect in the function of Forms with respect to accounting for the reality of numbers.

Phase 1: "All these views share a difficulty which occurs with regard to species-of-a-genus, when one posits the universals, viz. whether it is animal-itself or something other than animal-itself that is in the particular animal. . . . For when we apprehend the unity in 2, or in general in a number, do we apprehend a thing-itself or something else?" (1085^a23-30)

Without hesitation, Aristotle prefaces his criticism with the claim that as a matter of principle, a species/genus relation is, and presumably must be, present in any such analysis. Then, positing the "universal" (in his sense of universal) under scrutiny, for example, the number 2, he asks what exactly we secure when we seek "the unity in 2." If a particular number occupies the species slot, then what does the corresponding genus provide that establishes the unique unity of that number? Thus, 2, as an individual number, has its own unity, and this unity makes the number 2 to be what it is as that number. But what exactly is this unity and how would it differ, as an instance of unity, from the unity that would make 3 and all other numbers what they are? If there is an element that establishes the unique identity of this (or any) number, how can the underlying source of this identity be recognized and described? It will be noted that the appeal to unity in this context is crucial to Aristotle's criticism.

Phase 2: Aristotle criticizes (citing the *Phaedo*) the position maintaining that Forms cause existence, being, and becoming (1080^a2-5):

> In the *Phaedo* the case is stated in this way—that the Forms are causes both of being and of becoming. Yet though the Forms exist, still things do not come into being, unless there is something to originate movement; and many other things come into being (e.g., a house or a ring) of which they [the Platonists] say there are no Forms. Clearly therefore even the things of which they say there are Ideas can both be and come into being owing to such causes as produce the things just mentioned, and not owing to the Forms.

Two criticisms are at work in this passage. First, even if the Forms do exist, an entity of a certain sort cannot come into being unless something causes or engenders such a fundamental dimension of movement. Aristotle's assumption is that a Form cannot function by itself as a causal agent but requires what Aristotle refers to in the *Physics* as an efficient cause. Here again, Aristotle presupposes the truth of his own position, this time an approach to causality as developed in another work. Second, some things exist without the need of Forms as sources of their existence; as a result, the implied criticism is that the Forms do not command comprehensive control in initiating the existence of things.

The distinction between being and becoming is metaphysically broad, but its relevance to a strictly mathematical context seems, at least initially, tenuous. The number 2 must be caused, but in view of the argument detailed in Phase 1, Aristotle would simply deny that the cause of the number 2 is participating in a Form duality. Furthermore, if we remain in the world of the *Phaedo*, then it could be argued that duality, as a Form, exists prior to (in some sense of prior) any instance of duality, including the number 2. In this respect, then, the number 2 "becomes" from a generative process originating, ultimately, from the Form duality. However, this approach to the problem only relocates the position under scrutiny so that it becomes subject to the first criticism of Phase 2, that is, that the Form duality (assuming such a reality exists) does not, by itself, initiate the motion requisite for producing anything, much less a reality with the seeming stability of a number. A later passage (Book N, 1092ᵃ2-5 clearly states the main point of this objection: "Those who say that existing things come from elements . . . should have first distinguished the senses in which one thing comes from another, and then said in which sense number comes from its first principles."

Also, from the Whiteheadian approach (introduced in Chapter 2) with its emphasis on process as an essential feature in describing mathematical realities, we might conjecture that in order for numbers to exist, it is necessary that entities of some sort *already exist* as a necessary condition so that a number accurately identifying the gathered set of those entities can "come into being." It may be noted that whether or not the entities in question are explicitly enumerated does not apparently affect the fact that x number of entities are present with or without being indicated by such numerical means. And is the existence of the number 2 dependent on the existence of a pair of entities, a and b? If so, then numbers do not exist unless things other than numbers exist. This kind of special dependency would therefore itself become noteworthy as a reality of significant metaphysical proportion.

Phase 3: This objection concerns one of the standard difficulties with the Platonic theory of Forms; Aristotle presents it in the context of mathematical entities (1084^{a}25-30):

> It is paradoxical also that there should be an Idea of 10, but not of 11, nor of the succeeding numbers. Again, there both are and come to be certain things of which there are no Forms; why, then, are there not Forms of them also? We infer that the Forms are not causes. Again, it is paradoxical if the number-series up to 10 is more of a real thing and a Form than 10 itself.

Aristotle repeats the point that certain things can "come to be" without having the original impetus for existence due to the Forms (implying that the Forms are not then, in point of metaphysical fact, sufficient to explain causally the existence of particulars). But new is the claim that paradox obtains for asserting that a Form exists for the number 10 but not also extending such existence claims for the next number in the series, 11, not to mention Forms for all succeeding numbers. How do we "close" the group, or set, of numbers in order to account for each element existing within that group? Paradox also emerges if it is asserted that a "number series up to 10" (1, 2, 3 . . .10) is "more of a real thing" with its own originating Form than 10, the highest single number in that sequence.

One aspect of this criticism is the standard problem of the extent of the Forms. How many Forms are there and how do we proceed in order to determine the extent of their metaphysical scope, given the plethora of particularity that surrounds us? The Platonist could counter with the standard appeal to *Republic* 596a [Shorey's translation]: "We are in the habit, I take it, of positing a single idea or form in the case of the various multiplicities to which we give the same name." Applying this principle to Aristotle's criticism is apparently straightforward: If 11 fish are gathered in a creel, then this multiplicity can be named by "11" as "the single name" accounting for the number of items under scrutiny. Such identification could stand as a rejoinder to the particular example Aristotle cites (thereby implying the existence of the Form for 11 even if that Form is not explicitly identified as such elsewhere in the dialogues). In a sense, however, this move is only a stopgap measure since the scope of the problem remains unresolved.

Among other concerns, the *Republic* passage seems to tie the existence of Forms to the fact of language with a degree of intimacy that appears to threaten the full autonomy of the Forms as a distinctive realm of reality. We would want to assert, presumably, that Forms exist whether or not language has evolved in ways that provide names for such realities. That said, it also seems to follow that mathematical realities will have corresponding Forms as long as these realities can be named or otherwise represented in a linguistic (or, perhaps, symbolic) format.

Units and Collection—Criticism B [Associativity/Inverse Element]

Aristotle's fundamental position denying that it is possible to account for the existence of particular numbers by participating in a Form is supplemented, from different directions, by a variety of arguments in *Metaphysics* Books M and N criticizing the notion of Platonic Form-numbers. His arguments aim at a position in which each number is determined to be what it is by a Form-number that is itself characterized by units. The basic strategy, given that a Form-number includes units, is to examine how the units are, or can be, associated with one another (1081^{a}1-5):

> First, then . . . let us inquire if the units are associable or inassociable, and if inassociable, in which of the two ways we distinguished. For it is possible that any unit is inassociable with any, and it is possible that those in the "2-itself" are inassociable with those in the "3-itself" and, generally, that those in each ideal number are inassociable with those in other ideal numbers.

This passage states one way to approach the position under scrutiny. Unlike the Forms introduced in the *Phaedo*, a Form-number includes "units" in order to generate numbers, whether considered as numbers existing by themselves or as instruments for counting and calculation. Thus, if "units" are associated with one another, then it seems fair to say that such association names a process by means of which two (or more) units are gathered together for a specific purpose—to constitute the uniformity of each and every number as that number.

Consider the number 2. Aristotle asserts as a point of departure (1084^{b}6) that "each of the units is part of the number as its matter, and the number acts as form." If this statement of equivalence is accepted, Aristotle draws the conclusion (1084^{b}20-23): "For the universal is one as form or substance, while the element is one as a part or as matter. For each of the two is in a sense one—in *truth* [italics in text] each of the two units exists potentially (at least if the number is a unity and not like a heap, that is if different numbers consist of differentiated units, as they [the Platonists] say)."

The image of a "heap" [σωρὸς] is vivid and metaphysically consequential. The argument contrasts the unity of a number with a heap. Heaps can just happen, or they can be more or less the result of teleological activity (a heap of raked leaves). In the position under critical scrutiny, a number has a certain set of "differentiated units," that is, two units in the number 2, three units in the number 3, and so on. Also, each single unit among these units is identical to any other unit within the confines of the class in which these units are gathered, a condition required so that the number can become what it is. Presumably then one kind of heap could include a number of elements that were identical in all respects to one another except for numerical difference. Thus, a heap of sand could include a large number of grains that were identical in all respects to one another—the identity element proper to each grain would thus be "member of a heap." But such uniformity could not function in this respect alone as a factor in determining individual numbers. How then to generate the "twoness" of the number two?

Consider that for the number 2, its "twoness" is a unity; however, according to the Form-number position, this number is constituted by two distinct units. As such, each unit (recall, each unit construed as a constitutive factor in the number 2) is "in a sense one" if considered by itself, as *this* unit. But the argument maintains that in this case, there would be no distinction between the unities of each construed unit. As a result, Aristotle introduces, under the fundamental cover of matter and form, the metaphysical dimension of potency, inferring that in order to maintain that each of the units in the number two exists, each (and they) can exist only *potentially* with respect to the necessity of each unit being differentiated from the other unit. Note again that Aristotle does not hesitate to introduce his own metaphysical position in order to bestow a degree of feasibility to the Platonic position.

Here is Julia Annas usefully summarizing the core of this argument in terms of its basic metaphysical elements—form, unity, matter, potentiality, and unit:

> Form is actuality; a number does actually have a formal principle of unity. Matter is rather the range of a thing's potentialities; so to say that there are 3 units [if 3 is the number under scrutiny] is just to say that we *can* [italics in text] count that unit three times. Units do not have actual existence; their existence is merely potential and is dependent on our actually counting. So, since for Aristotle what is actual is always prior to what is merely potential, the existence of the number as a unity is prior to the existence of the units in it. (JA, 182–3)

It follows then that an appeal to units as the cause of the existence of a number fails, since the number under scrutiny *already exists* prior to the existence of the units in that number, which putatively and, for Aristotle, only potentially account for that number's existence.

In closing, Annas remarks (JA, 183) "Aristotle's philosophical terminology is somewhat unilluminating when applied to numbers and units." Annas' studied hesitations regarding the apparently "unilluminating" character of Aristotle's language are not developed to the extent that they might be in terms of the underlying metaphysical positions. The force of Aristotle's criticism is to approach the three units in 3 and then to take their asserted existence as actual units and reduce that mode of existence to something very different, that is, potentiality as a sort of inverse of actuality. Since form is prior to matter, as Annas brings out, there must already be a form of 3 existing in order to establish the possibility that the putative three units constituting the Form-number of 3 (or "threeness") can be counted as the triple entities supposedly making 3 to be the number it is. But since the same entity cannot be both in potency and in actuality at the same time and in the same respect, the unity of a number, 3—that is, the form of that number—cannot be said to consist of a collection of 3 distinguishably distinct units (even if the "threeness" is produced by counting just a single unit three times). In sum (as stated above), the argument shows that a number, when taken as a unity (by reason of its form), must exist prior to the existence of the units purportedly constituting that number; if so, introducing units for purposes of generating a number becomes vacuous and unavailing.

For Aristotle, we cannot collect the constitutive elements of a given group as if they were a "heap" of undifferentiated realities. More precisely, if we attempt to group two or three entities in a set and then relocate that grouping (as associativity, a condition of group theory, allows), we are concealing the metaphysical fact that each such entity cannot merely be identified as this or that unity ("a" or "b" or "c"), but that each of these apparent unities must be analyzed in order to determine what kind of existence it possesses. Thus, the parentheses that partially define associativity as a formal principle help focus attention on what elements so grouped have in common when viewed from a higher level of abstraction. The identity (a + b) = (b + a) contributes heuristically to the realization that although the two units can be arranged in this way, they are not identical with respect to their conjunction, as a supposed unity within another unity, that is, the number in itself. For Aristotle, both a and b (regardless of the order of their conjunction) exist on a different level of reality. Aristotle's basic principles require that the elements of a supposed "heap" must be determined according to a fundamental metaphysical descriptor (i.e., in his metaphysics, potency) *prior to* any kind of analysis or identification of that collection of entities. The unity of potentially existing entities in effect functions as a replica of the inverse element in group theory, that is, it undercuts the possibility that the entities appealed to can exist productively, in a positive way, with respect to the existence of the unity (the number itself, whatever it may be) they purportedly generate. We note again the crucial importance of various dimensions of unity in the development of this reasoning.

A second argument (1082^a1-11) with a more intricate strategy is also classifiable, according to its basic components, as pertaining to aspects of grouping units as a factor in the generation of numbers. The argument is quoted verbatim (in the Annas translation). Since Aristotle's reasoning is tight and complex, we subdivide the text of the argument into four steps and follow each step with a comment. These comments include Annas' restatement and clarification when helpful; the discussions will also introduce elements so that the direction of the argument leads more readily into parallel positions in the metaphysics embedded in the Platonic doctrine of collection and division. As noted in the Overview, in Chapter 6 we will apply Aristotle's criticisms of mathematical positions to the metaphysical dimensions of collection and division.

Step 1. For example, in the original Ten there are ten units, and ten is composed both of these and of two fives. Since the original Ten is not just any number and is not composed of just any fives (or just any units), the units in this ten must differ.

Comment: The number ten (the "original Ten") is unique, as are all numbers. There are ten units "in" the number ten in order to make it this number and not any other number. The claim that ten "is not composed of just any fives" asserts that the two groups of five units must be unique as groups. Annas makes the point this way (JA, 171): "The argument runs: 10 is made up of two 5s; but since 10 is a special sort of number, so will be the numbers making up that number, and so these 5s, each of which makes up a special 5: if they did not thus fall apart, there would not be two 5s in 10."

The number 10 is a "special sort of number" because each and every number is "special"; the number-Forms 2 and 10 are as different from one another as, for example, the Forms Beauty and Justice. Furthermore, the property of number uniqueness also holds for each number "making up" a given number. Since the number 10 is made up of two 5s, Aristotle reasons that each of these groups is different and unique, qua "5ness" (so to speak), as the other group.

The factor of associativity is present in this reasoning, although the number of elements being associated with one another is five (and not two). The relevant principle at work is that individual elements of a more inclusive expression can be examined by grouping them in such a way that the subsets so grouped can be compared with one another, and also that inferences can be drawn from such comparison. Thus, in the example under scrutiny, if there were not two 5s (i.e., two groups each containing five units) in 10, there would not be 10 (given the arithmetical fact that $5 + 5 = 10$).

Step 2. If they do not differ, the fives of which the ten consists will not differ either, but since they do differ, the units will differ too.

Comment: Annas' rendering of this step appears to impute to Aristotle's reasoning the fallacy of division, that is, that what is true of the whole is true of each part of the whole. However, Aristotle is merely citing a theoretical necessity that a group of units constituting a number (5 in this case) will and must have a certain configuration of units. Strictly speaking then, Aristotle is merely pointing out a necessity pertaining to a part of a whole which would apply to that part even if it, the part, were considered as itself a whole, that is, if we were merely considering the number 5 and describing that number's metaphysical constitution. As such, this step seems more clarificatory than consequential; however, the argument now shows the strategic relevance of this step.

Step 3. But if they do differ, will there be no other five in <the ten> but only these two, or will there be?

Comment: Annas observes (JA, 171):

> If there are two *different* 5s in 10 then there is some relevant difference between the two groups of 5. . . . But then there is no guarantee that this is the *only* relevant difference; given 10 units, there are many ways (252 in fact) of dividing them into two 5s. Merely picking out two groups of 5 does not entitle us to say that we have picked out *the* two fives in 10; we can always at least ask . . . whether there are other 5s that can be put together from other combinations. [all italics in text]

In sum, if other combinations of five elements are possible, then we have not in fact accounted for 10 by pointing to *this* and *that* set of five elements. An appeal to the "5ness" of any two sets of five things does not constitute the uniqueness of these two sets with respect to the generation of ten as itself a unique number constituted, as the Platonists maintain, from sets of units.

Step 4. It is absurd if there are not; but if there are, what kind of ten will be composed of them? There is no other ten in the ten over and above itself.

Comment: But if *different* groups of five things can, when added together, form a supposedly unique number, ten, what kind of ten is this? On the strength of the principle that the constituting units must differ among themselves, the result of combining *any* two groups of five units will be multiple, therefore generating in some sense *different* kinds of 10. But since each number is unique (by definition), there can be only a single number 10. Thus, for Aristotle, the possibility of grouping *different* subsets of 10, each of which contains 5 units, is inconsistent with the claim that such grouping can be employed to generate *one and the same* unique number.

The way in which grouping applies to elements in a subset of a whole makes a significant difference once we specify what kind of whole (with respect to its unity) is under scrutiny. It may also be observed that the argument has introduced three distinct senses of unity: the unity of a unit, the unity of a subset composed of units, and the unity of the whole composed by the conjunction of constitutive subsets.

Units and Predication—Criticism C [Identity Property]

Predication is an important contemporary concept in metaphysics as well as in the philosophy of language. One of its earliest appearances is in the *Phaedo*. At 102b, Phaedo is recounting Socrates' announcement of the Forms: "As I recall it, when the above had been accepted, and it was agreed that each of the Forms existed, and that other things acquired their name by having a share in them." Thus, things are named and described based on the Forms in which those things participate. Predication involves language, a factor that is clearly relevant to any metaphysical inquiry (especially in the contemporary philosophical world). But in the sense in which it will be discussed here, it is important to appreciate that predication—to say something of something else—is also decidedly metaphysical in significance and structure. Thus, accounting for the referential function of language depends, ultimately, on metaphysical considerations.

We will therefore approach predication in an equally broad sense, with an emphasis on predication in expressly metaphysical contexts. What can we say of something if we want to identify that something with respect to more inclusive, general realities? We shall see that the Abelian factor of association serves as an incisive methodological gambit for concentrating attention on a limited sub-grouping of elements, thereby facilitating the discovery of substantive differences emerging from the elements so connected. In parallel fashion, if we approach collections of units intended to establish conclusions in the philosophy of mathematics, the identity element taken as a heuristic pointer will facilitate indicating what properties, and kinds of properties, can be properly predicated of a particular thing with respect to its identity.

Here is the first argument involving predication we shall examine. Aristotle's reasoning is aimed at the notion of units and how they are understood when considered the defining element of numbers. The discussion, as we shall see, evokes a range of senses of unity, all of which are relevant to metaphysical considerations (1084^{b}30-2):

> Therefore the unit becomes the matter of numbers and at the same time prior to 2; and again posterior, 2 being treated as a whole, a unity, and a form. But . . . because they [the Platonists] were seeking the universal they treated the unity which can be predicated of a number, as in this sense also a part of the number. But these characteristics cannot belong at the same time to the same thing.

As a point of departure, consider Ross' note (Vol II, 906, fn49) on this passage: the Platonists "treated the unity which is predicable of a number, as well as the unit in a number, as a part of the number." For Aristotle, this way of approaching the structure of a number illicitly combines elements that embody three separate approaches to a global understanding of a particular and therefore must be kept apart. The argument assumes that the proponents of this approach were, again, seeking a "universal," again in the proper Aristotelian sense. If we say "two" with the intent of referring to the number 2, then "two" is a unity since it is what is predicated of a single number. But the Platonists contend that the unity of the number 2, as a whole, is constituted by units unique to the number 2. However, this attempt conflates, from the part/whole perspective on an individual thing, what makes the number 2 to be what it is, its parts as units, with what that number is as such, as a whole, that is, a unique number. In view of what does the number 2 possess its identity as that number and no other? In sum, the position under scrutiny fails to distinguish properly between parts and wholes in its attempt to render a metaphysical account of a number. The unity of a part of a whole differs from the unity of the whole in which that part exists.

A second argument (1083^{a}1-2), introducing additional elements borrowed from Aristotle's own metaphysics: "First of all it is well to determine what is the differentia of a number—and of a unit, if it has a differentia. Units must differ either in quantity or in quality; and neither of these seems to be possible." An Aristotelian appeal to "differentia" presupposes the species-genus format, since a differentia (if intended in the technical sense) establishes a property which, when predicated, will distinguish one species from another species when both types exist in the same genus. Aristotle again assumes his own metaphysical categorical scheme in this criticism. However, the criticism retains its force even if we hold in abeyance Aristotle's basic metaphysical scheme; the relevant point becomes whether it is possible to say anything in an attributive sense about the units. Or are they some variation on, to use contemporary modes of designation, "bare particulars" or "irreducible particulars"? Can a unit in the sense employed by the Platonists in the context of determining the uniqueness of individual numbers simply be a kind of unity without any additional property specification?

When Aristotle contends that units "must differ either in quantity or quality," the modality of "must," if taken as a kind of metaphysical necessity, suggests (if not implies) that if a unit lacks specification in both categories, it becomes doubtful whether we are discoursing about anything existent at all. The implication is that Aristotle will not countenance the possibility that a unit could exist unless it exhibited predication as specified by categories of this sort.[4]

A third argument, with a more generic criticism along the same lines, appears after an iteration of the Aristotelian axiom that being has many senses (1089ᵃ7). For Aristotle, it follows that 1089ᵃ12-14) "It is absurd, or rather impossible, for a single kind of thing to become responsible for one thing's being this, another being such, another's being so much, another's being somewhere."

Aristotle's introduction of "being such," "being so much," "being somewhere" may be taken as *categorical* specifications. If so, then the objection is that a "single kind of thing" cannot generate multiply distinct categorical properties in the same "one thing." The force of the argument depends on assuming that (a) the single kind of thing is a Form or a Form-number and that (b) these realities do not have the requisite metaphysical diversity in themselves to endow any "one thing" with all the properties which this one thing may possess (nor with a multiplicity of different things of the same type having different properties when the things are compared with one another). The predication of the categories of quantity and place seems clear from the wording of the claim.

A fourth and final (for our purposes in this context) argument shows Aristotle's unflagging attempts to penetrate the reality of a unit as the Platonists have employed this notion in mathematical matters. The concern for the connection between the unit and unity is evident in the formulation of the criticism (1085^{b}11-16):

> Above all one might press the question "if each unit is one, what does it come from?" Certainly each is not the one-itself. It must, then, come from the one-itself and plurality, or a part of plurality. To say that the unit is a plurality is impossible, for it is indivisible; and to generate it from a part of plurality involves many other objections.

This passage indicates that for Aristotle, the Platonic unit has at least one discernible property, that is, it is indivisible (as is a point on a line). If so, then a kind of unity that is also indivisible (according to Aristotle) cannot be construed as a plurality since the contradiction is immediate. This reasoning reinforces the need to investigate the notion of unit in order to determine, if possible, how it should be understood in general, metaphysical terms.

Units and Sequence—Criticism D [Commutativity]

In Book N, Aristotle initiates a new line of criticism by, first, registering a general complaint (1090^{b}14-20–Annas' translation): "Besides, if one is not too complacent, one might go on asking questions about all <kinds of> number and mathematical objects, and the way they make no contribution to one another, the ones coming before to the ones coming after. . . . However, judging by what we can observe, nature is not a mere series of episodes, like a bad tragedy." The argumentation in Criticism C has shown that it is not theoretically possible to predicate any property of a number if its existence depends on units, that is, a relation between the unit (or sets of units) and a characteristic other than itself (and also constituted by units) cannot be drawn. But it is important to appreciate that numbers exist in additional, more specific kinds of relations with respect to one another.

Consider the sequence (the second argument, discussed above in Criticism B) of the numbers from 1 through 10. This set of numbers follows a definite sequence—1, 2, 3 . . .10. If any number gets out of order in this sequence, the sequence is destroyed. Thus, a sequence of this sort is itself a unity such that each unit after the first unit can be, in Aristotle's words (1081^b29) "prior and posterior." This "sequential unity," as it might be dubbed, is real and must be accounted for as far as the modes of reality of individual numbers are concerned, insofar as individual numbers collectively and in ordered manner constitute this grouped unity (1 to 10 taken as a single sequence) as well as potentially limitless additional grouped unities of this sort.

Consider the factor of commutativity, a necessary feature of an Abelian group. If the numbers are construed as groups of units and units as such are understood to be completely abstract and undifferentiated in any way, then $(a + b) = (b + a)$ lays out a condition that allows ready recognition that if the plus sign is replaced with the "is greater than" sign, the resulting identity is manifestly false and indeed self-destructive to the reality of the numbers so deployed. It should be observed then that critically construing a mathematical phenomenon from the standpoint of this particular condition of an Abelian group illustrates again the useful heuristic value of aspects of this group's structural complexity when one feature of that structure is confronted with mathematical realities approached from broader, more abstract perspectives (i.e., $a > b$ rather than $a + b$).

It is readily possible to extend this line of criticism incumbent on the unity of a given number, and in two distinct dimensions—even/odd, and location in a sequence. If the units in a given number are not comparable with the units of any other number, then there is no obvious way to situate the existence of any one number regarding its unity as that number (a) with respect to priority and posteriority in relation to the existence of any other number. Thus, a critical appraisal of numbers according to the conditions of commutativity and the identity element reveals the need for some kind of reality that will uniformly change a sequence of numbers so that the successor of any given number exemplifies the other type of a pair of determinately opposed predicates (i.e., odd/even). In addition, this sought-for reality must guarantee that a given number will (b) retain its unity but also can become situated in a sequence including other numbers. It will be noted that the prior/posterior relation is more inclusive than the sequential location since, for example, 4 is posterior to 2 and 3 but prior to 5 . . .10, yet 4 occupies one and only one location in the sequence of numbers from 1 to 10. Obviously, both phases of relationality must be satisfied for any given number past 1 (assuming, as we will here, that 1 is itself a number). The factor of commutativity, fundamental to Abelian group theory, also provides a formal condition that, when applied to the behavior of units in the Platonic theory, allows the necessity of numbers being involved relationally and in sequence with other numbers to become apparent, thereby raising relevant questions concerning the cogency of units as the fundamental elements of numbers.

Division and Types—Criticism E [Associativity, Commutativity]

Forms stand as the source of predication and also cause types of things; we may wonder whether Form-numbers possess similar functions. In the final chapter of Book N, Aristotle examines whether numbers can be construed as causal agents. Thus: "There are seven vowels, the scale consists of seven strings, the Pleiades are seven, at seven animals lose their teeth (at least some do, though some do not), and the champions who fought against Thebes were seven." Aristotle asks: are these numerical identities, occurring in widely disparate settings, *caused* by the number seven "because the number is the kind of number that it is . . .?" (1093^{a}14-18)

As we might expect, the answer is no. Aristotle introduces a distinction regarding types of reality in order to justify his denial of anything like causation present in this collection of entities characterized by the commonality of the number seven. Here is the argument (a different example, that is, the number of seasons in the year, illustrates his point):

> For the seasons and a particular kind of number go together; and the other agreements that they [the Platonists] collect from the theorems of mathematics all have this meaning. Hence they are like coincidences. For they are accidents, but the things that agree are all appropriate to one another, and one by analogy. For in each category of being an analogous term is found. (1093^b13–18)

The number seven is identical to all the members of a given set. But Aristotle says this identity is accidental to each of those members. The accidental character of a number generates a kind of analogical unity as we move between and among the various members of the group. All vowels are identical to one another as vowels, but the fact that there are seven such realities is an accidental feature of the grouping of these elements. In the same way, each string of a scale is identical to every other string as an element in a scale; the fact that there are seven such strings is accidental to the reality of the strings insofar as they are, individually and collectively, members of a scale.

Associativity and commutativity both contribute to establishing connections between and among the items of each group of 7, thereby showing a uniformity that applies collectively to all the items of each 7-member group ($a + b = b + a$; $b + c = c + b$, etc.). This process will facilitate recognition of identities present in any two (or more than two) members of each group. However, the unity of 7 insofar as it pertains to multiple distinct sets of items, is, for Aristotle, strictly analogous. The fact that 7 applies to multiple sets of distinct items has nothing to do with the content of each of the sets identified in the argument. Furthermore, the set of seven contains items belonging to different categories of being, and this metaphysical fact is sufficiently foundational to preclude construing the identical number common to all the items in each set as anything more than an accidental feature of the set taken as a collected unity.

A second, more intricate argument has a direct bearing on the possibility of distinguishing types of being and identifying determinate classes. This argument depends to a considerable extent on the notion of "measure" and the approach Aristotle takes toward this notion and its connection to unity. Consider then that Chapter 1 of *Metaphysics* Book N opens with an analysis of principles deployed as contraries and includes Aristotle's adamant rejection of this strategy as far as producing an adequate account of the generation of numbers. In the course of this discussion, almost as a parenthesis, he asserts that "a real object has no contrary" (1087^{b}1). The number one may appear to illustrate this claim; however, Aristotle contends that "one is not a real object in its own right." In fact, it is obvious that "one means a measure [μέτρον]" and that "in every case there is something else that is the subject"; thus, the number one "means a measure of some plurality, and number means a measured plurality and a plurality of measures." For Aristotle, "there is good reason for one not to be a number: a measure is not itself measures. The measure and one are both principles" (1088^{a}5–7). And, as principles, both one and measure are, for Aristotle, different sorts of reality than particular objects.

Aristotle's denial that one is a number will be discussed in Chapter 6 (since, as the reader will observe, it conflicts with the argument supporting the status of one as a number presented in Chapter 1). For now, it is important to recognize that in Chapter 1 of Book N, Aristotle presupposes this approach to the number one and its application to singular entities. Aristotle's complete argument follows (1088^{a}7-14). The content is compressed and also important with respect to Aristotle's understanding of number and unity. (The argument is also, as we shall see, incisively related to aspects of collection and division.) For the sake of clarity, it will be useful to set off the various steps. The subsequent discussions, numbered 1 to 4, are based on the corresponding enumeration of Aristotle's text (Annas' translation):

[1]. The measure must always be some one and the same thing applying to all cases; for example, if there are horses the measure is horse, if men it is man.
[2]. If there are a man, a horse, and a god, the measure will perhaps be living thing, and their number will be a number of living things.
[3]. If there are a man, white, and walking, they will hardly have a number, because they all belong to the same thing which is numerically one.
[4]. Still, they will have a number of categories of some such term.

In general, two distinct types of unity are at work here: First, the unity of each of three independently existing elements; second, the numerical unity of a single element, but an element subject to predication from three (at least) different perspectives, with different properties produced by each of these perspectives.

The following commentary brings out in more detail how these senses of unity function in Aristotle's reasoning:

Step 1: There are pluralities of particulars under analysis, and measure represents an identity predicable of all instances of a given group. We are not told whether measure, as such, can be taken to represent a kind of reality that may differ, as a kind of unity, from the unity exhibited by each individual instance of a given measure.

Step 2: If the group is composed of three different things, then it is possible that a measure can be specified which covers all three things. Note that in Aristotle's example, the three things significantly differ among themselves (especially if we analyze the set of three according to the dyads available from the application of associativity). Thus, if we combine man and horse and then examine this conjunction, we recognize two different kinds of things; if we combine horse and god, the same basic recognition of difference also follows, and so on Aristotle's guarded use of "perhaps" suggests a degree of hesitancy concerning the legitimacy of, for example, "living thing" as a justified measure. (We may be permitted to wonder why "living thing" is not readily granted as a measure in the full canonic sense of the term as it is being presently employed. Aristotle's metaphysical instincts might suspect that the concreteness of each individual in this group may be somehow compromised if its unity as an individual of a specific type of thing is subsumed under a reality—living thing—of such considerable generality.)

Step 3: Note the indirect importance of number. In the third step of the argument, it appears that each item listed can be assigned the number one (one man, one instance of white, one instance of walking), and it will follow that the set under scrutiny has one plus one plus one = three members. But Aristotle denies that such an enumeration is possible. Thus, each member does not for Aristotle exist as an independent singleton; rather, each unit is a unity by virtue of being the subject of predication based on something else. Also, each unit is apparently a different kind of entity when juxtaposed with its neighboring unit. The functions of addition as well as commutativity usefully provide a formal basis for drawing this conclusion. Aristotle has distinguished two aspects of one substance (a person) and then denies that the resulting set of three elements will have a number since all three belong to one thing, the person as such, that is, as a unity which is "numerically one." The substantial dimension of the entity under scrutiny (the single person) effectively dissolves any process of individuating the entity in any sense. At this level of analysis, one person is just one person, however many distinguishable attributes belong to that single entity.

Step 4: Despite the conclusion just reached, Aristotle nonetheless admits that "categories" will be applicable to each of the three items. This admission is of considerable interest, since it appears to sanction the following argument—if different categories can apply to each of the three items, then surely there are differences in the kinds of categories applicable depending on the identity of the item. Thus, a category that situates "white" as an instance of color may not cohere with the item of "walking" even if we grant, as Aristotle seems to do here, that both white and walking exist in ways that admit the possibility of categorical predication. Given the very possibility of employing categories, which of course is fundamental to Aristotelian metaphysics, it seems necessary to admit that each item has some sort of determinable unity; for if such unity did not obtain, then how could any category be applied to something that does not have a sufficient mode of existence by itself to support the predication of categorical existence? Worth noting for possible future relevance is that Aristotle's position appears to rest on a decidedly stringent understanding of unity and its possible manifestations in countable things, especially with respect to the predication of categories as vehicles of differentiating and describing countable things.

As noted at the outset of this chapter, the arguments laid out and analyzed in Chapter 5 represent only a selection of the criticisms Aristotle brings to bear against the Platonic philosophy of mathematics. These arguments illustrate the conceptual breadth and range of Aristotelian criticisms (and they also, as we shall see, cohere with and reinforce the general relevance of the argument developed in Chapter 6). A close examination of *Metaphysics* Books M and N in their entirety would reveal an extensive variety of analytical approaches to what appear to be inherently dense dimensions of reality and, as a result, dimensions offering wide-ranging theoretical possibilities for metaphysicians of sufficient courage to try their hand at their art. But for Aristotle himself, the onslaught of arguments developed in these two books is sufficient as far as dealing adequately with the putative cogency of the Platonic approach to these mathematical matters.

Aristotle's concluding position regarding the Platonic postulation of units as the basic constituent element of numbers and how units may be associated (or not) with one another is summarized in Chapter 7 of Book M (1082^{b}3-8):

> In general, to differentiate the units in any way is an absurdity and a fiction; and by a fiction I mean a forced statement made to suit a hypothesis. Nor neither in quantity nor in quality do we see unit different from unit, and number must be either equal or unequal—all number but especially that which consists of abstract units—so that if one number is neither greater nor less than another, it is equal to it; but things that are equal and in no wise differentiated we take to be the same when we are speaking of numbers.

Aristotle again appeals to two of his standard categories, quantity and quality, as linchpins justifying the denial that units lacking any kind of differentiation according to these categories can supposedly constitute a number. In fact, Aristotle generalizes this claim, stating it with considerable force when he asserts at the outset that to differentiate units "in any way," presumably by appeals to a differentia of any kind, is "an absurdity and a fiction," nothing more than an *ad hoc* device to support a prior claim concerning the attempt to account for the reality of numbers. This reasoning is then extended into a criticism based on the lack of an identity factor. Numbers are either equal to one another or not equal to one another. But if it is impossible to determine, of any two numbers, whether either is greater or less than the other and thus are "in no wise differentiated" from one another, then these two numbers—asserted to be different "we take to be the same." We may detect in this reasoning an anticipation of the identity of indiscernibles. If two "abstract units" are completely and in all respects abstract, then for Aristotle we are virtually compelled to conclude that we are not dealing with two items at all, that is, the two ostensibly different numbers (claimed to be differentiated by completely undifferentiated abstract units) are, in point of metaphysical fact, the same number.

Later in Book M, Aristotle consolidates his critical position, drawing an important conclusion concerning number and Platonic attempts to examine and disclose its metaphysical makeup in order to declare number to be a separately existing entity (1085^{b}34–5): "All these objections, then, and others of the sort make it evident that number and spatial magnitudes cannot exist apart from things." As the point might be stated, for Aristotle, things and numbers are metaphysically coextensive, with a dependency relation of sorts connecting the existence of numbers with the existence of things numbered. The final sentence in Book N, and thus the final sentence of the *Metaphysics*, also epitomizes this conclusion: "The fact that our opponents have much trouble with the generation of numbers and can in no way make a system of them seems to indicate that the objects of mathematics are not separable from sensible things" (1903^{b}27). The appeal to the apparent need for a system indicates that for Aristotle, an account of any one sector of reality, for example, the generation of numbers and mathematics in general, must cohere with accounts of some if not all other sectors of reality. The implication is that claims about mathematics, first, must be metaphysical—that is, broadly based in content—and second, must be developed so that this content pertains to other types of reality and does so according to systemic conditions. This conclusion, if accepted, sanctions the following section as a fit and forward-looking conclusion to this chapter.

Critique and Pattern: A Metaphysical Mosaic

A summary reaction if we were to review all the arguments that appear in Books M and N of the *Metaphysics*: From a standpoint of discursive precision, Aristotle's critical treatment of Platonic philosophy of mathematics in general, and number in particular, stands as somewhat of a mixed bag of approaches—from one-line rejections to arguments incorporating one or two premises (usually curtly stated), to more expanded and at times elegantly developed exercises in reasoning executed in something approaching formal rigor, all of this variety inflected with argumentative snippets laid out here and there in something less than pristine syllogistic order. And as noted at the outset of this chapter, the arguments examined above represent only a selection of the criticisms Aristotle deploys in *Metaphysics* Books M and N. This kind of polyglot philosophizing occurs more than once in Aristotle's texts and, as such, presents a range of interpretive decisions as far as its relevance to individual positions of other philosophers is concerned. If, however, we allow these exercises in critical assertion and reasoned argument to exert their intent at assessment as fully as possible, we will recognize (as Julia Annas did in her commentary) that all things considered (and duly acknowledging that our texts of these works are surely edited by others), they are penetrating as well as diverse in strategically elucidating and evaluating Platonic philosophy of mathematics.

Consider also that if we review the arguments discussed in Chapter 5 for the purposes of identifying their pivotal notions, we may list the following elements:

> species/genus, universal, particular, participation (as a type of technical relation), unit, combination, uniqueness, potentiality, prior/posterior, part/whole, identity, difference, sequence, measure, associable, similarity, one, form, property, relation, hierarchy (i.e., degrees of reality), substance, quality, quantity, subject, contrary, "heap" [distinct entities amassed but jumbled together], nature, differentia.

In a word, the "stuff" of metaphysics! For Aristotle, an appropriately philosophical analysis of mathematical considerations does, and doubtless must, operate with the elements, mechanisms, and argumentative rhythms of metaphysical analysis. And if we take a long view from a historical, present-day perspective, it is fair to assert that the various elements in this compilation in fact animate much of the metaphysical philosophizing that defines the Western tradition.

We are, perhaps, struck by this broad-based array of critical approaches, the diversity of technical vocabulary, and from an overall narrative standpoint, the fact that the concluding two books of a major work in metaphysics are devoted to what may initially appear to be an adjunct concern as far as the comprehensive scope of the metaphysical enterprise is concerned.

This arguably dismissive attitude toward mathematics in the context of metaphysical inquiry is, however, misplaced. Indeed, the next chapter demonstrates that when we deploy a cross-section of the metaphysical elements just listed in service to an alternate set of basic principles—they are Platonic but in contexts differing widely from those circumscribed by mathematics—the various (and multi-various) possible reconfigurations and developments involving sets and subsets of these metaphysical elements become large and diverse.

Chapter 6 will demonstrate that the relevance and resiliency of the elements driving Aristotle's mathematical critique here in Chapter 5 provide clarification and critical insights into the theoretical dimensions of the Platonic position on collection and division. Indeed, if we adopt a certain inclusively motivated interpretive perspective, it becomes even more evident that the mathematician and the metaphysician are working with the same basic elements. If so, then it would not be out of place to anticipate that an application of Aristotle's concentrated work on Platonic mathematics will also be relevant, perhaps incisively so, to Platonic metaphysics.

Notes

1 Julia Annas lists the major arguments on unity which appeared earlier, in *Metaphysics* A (JA, 16–17). For a detailed, meticulously argued discussion of this phase of Aristotle on unity, see Chapter 2 "The Ways of Being One" in Edward C. Halper, *One and Many in Aristotle's Metaphysics* (Las Vegas: Parmenides Publishing, 2009), 53–152.

2 *Aristotle's Metaphysics*, ed. W. D. Ross, 2 Vols. (Oxford: Clarendon Press, 1)924. Vol. 1, lii. All subsequent translations from the *Metaphysics* are by W. D. Ross in *The Basic Works of Aristotle*, ed. Richard McKeon (New York: Random House, 1941). Other translations will be indicated.

3 For a general account of Aristotle and mathematics, see Jonathan Lear, "Aristotle's Philosophy of Mathematics," *The Philosophical Review* 91, no. 2 (1982): 161–92. See also A. Graeser, ed., *Mathematics and Metaphysics in Aristotle* (Bern: Paul Hapt, 1987). For a standard critical analysis of Aristotle and the existence of

Form-numbers, see: Harold Cherniss, *The Riddle of the Early Academy* (Berkeley: University of California Press, 1945).

4 Ross comments concerning 1089^{b}35 (Vol. II, 477): "The manuscript reading seems to mean that if the unit is not some finite measure it is the limiting case of quantity–infinitesimal quantity; it is always quantity of some kind." In other words, according to this manuscript reading, it is a necessary but not sufficient condition for a unit that it have quantity, which presumably presupposes that a unit must be tangible in some sense.

6

Aristotle and Grouping

An Intersection of Mathematics and Metaphysics

The sub-title broadly describes this chapter's basic intent. Aristotle has criticized a Platonic (or Platonically-inspired) series of positions on basic issues in the philosophy of mathematics. In this process, Aristotle employed a variety of notions which, as noted at the conclusion of Chapter 5, eventually helped define the tradition of Western metaphysics. These abstract notions are extremely malleable in their applicability to theoretical accounts of reality. However, the ways in which the conceptual structure of mathematics play into underlying metaphysical concerns may not be so obvious. In this chapter, we shall explore an intersection of mathematics and metaphysics—Aristotelian elements of analysis initially animating a critical examination of mathematical issues but here juxtaposed with a later Platonic metaphysical venture.

Primary Objectives

These introductory remarks crystallize the results of the investigations in Chapter 5 and provide sharper detail concerning the strategy of the criticisms levied against Platonic positions concerning number. With the import of this foundational statement in hand, we will more readily appreciate the extent to which Aristotle's lines of argument can apply to Plato's position on collection and division.

Group theory is only one branch of mathematics. However, the notion of a group, that is, a set of conditions all of which are necessary to constitute a theoretical unity, as well as the processes involved in producing the group, elevates its philosophical relevance as a foundational and pivotal form of mathematical endeavor exhibiting a variety of interrelated realities. The structure of an Abelian group presents a diversely varied ensemble of factors, all located under a systemic and rigorous unity of functions. From this perspective, an Abelian group becomes a prism of sorts, with a structure such that through it, we may examine abstract but parallel configurations aimed at producing purely metaphysical ends.

Chapter 5 reviewed a representative sample of Aristotelian criticisms of a variety of fundamental positions in the Platonic philosophy of mathematics. The principles used in these critiques were similar if not identical to the underlying conceptual elements that constitute the structure of Abelian groups. Now it was demonstrated (in Chapter 4) that a series of affinities link Plato's methodology of collection and division with the structure of Abelian groups. Here metaphysics and mathematics intersect at a foundational level. Furthermore, it will become evident from the results of the analyses in Chapter 6 that the same principles used as critical wedges on Platonic mathematical philosophy can also be incisively applied to Platonic metaphysics (as embodied in applications of the methodology of collection and division). An inference of some interest emerges: this array of positive results is due in large part to the fact that the arguments analyzed in both Chapters 5 and 6 are based on *common abstract notions*—with divergent senses of unity often in the vanguard of the commonalities at work. If so, we may wonder whether this conclusion can be generalized to encompass additional areas of mathematics other than and in addition to the systematic insights provided by the applied structure of Abelian groups.

In sum then, Chapter 6 extends the scope of the criticisms introduced in Chapter 5 so that we are able to recognize the relevance of these criticisms for evaluating the conceptual structure of collection and division. Aristotle's criticisms, considered as a kind of organic analytical whole, become an array of critical approaches for appreciating a methodology aimed at producing metaphysical results; these criticisms therefore extend *from* evaluative treatment of theoretical considerations directed toward predominantly mathematical concerns *to* assessing a more inclusive exercise in speculative metaphysics. The resulting investigations will have various outcomes—sometimes critical in a negative sense, sometimes clarifying the import of the passages concerning collection and division, and sometimes heuristically illuminating possible developments of the position as originally stated. Concisely and summarily stated then, Aristotelian criticisms, initially introduced in mathematical contexts, have significant metaphysical implications. A cautionary note: this configuration of passages and their deployment in the thrust and counter-thrust of philosophical exchange does not constitute a preliminary phase of a knock-down confrontation between Plato and Aristotle. We are only exploring the range of discussion and relevant questions arising from the juxtaposition of theoretical work drawn from two important metaphysicians when filtered through positions on fundamental mathematical elements.

Given the affinity between Abelian groups and collection and division, we should perhaps ask whether all five elements in the definition of an Abelian group will generate equally effective counterpart repercussions when they are consonant with critical treatment of constitutive factors of collection and division. When aspects drawn from this area in mathematics intersect at a high level of generality with a metaphysical position, a full complement of consequences is not always to be expected. As long as an application of the defining elements of Abelian groups elicits relevant (and in some cases previously unrecognized) consequences and interpretive issues—especially when these results are theoretically or logically interrelated—the student of metaphysics may and perhaps should realize that this kind of juxtaposition between mathematical and metaphysical elements represents an integral approach to heightening awareness of the notional depth of certain jointly functional metaphysical structures.

Procedural Prelude

The discussions to follow explore the process of collection and division as if a Platonic metaphysician were attempting to put this systematic approach into practice. We begin with collection, then proceed to division. The sequence of themes is an interpretation derived from (a) the passages detailing collection and division discussed earlier in the book and then (b) deployed according to structures common to various modes of metaphysical thought. Each discussion cites relevant criticisms as developed in Chapter 5; also identified are various defining conditions of Abelian groups when they are instrumental to the stated lines of criticism. Finally, the discussions employ asterisks; an asterisk marks the appearance of certain terms of wide generality. These terms, employed at pivotal junctures in the various discussions and critical comments constituting Chapter 6, will be extracted, gathered, configured, and discussed in the systematic theorizing developed below, Chapters 7 and 8.

Collection: A Procedural Outline

For many metaphysicians, immediate reality is typically viewed as a bunch (or heap) of particulars. If the metaphysician is to instill order in this bunch, some rules or procedures seem necessary. We perceive a plurality of particulars. Now what?

From a Platonic standpoint (and stated in summary fashion), we collect things, then divide them according to the active agency of Forms. As sketched in the dialogues, collection is an organic process; the following enumerated list registers one way to proceed with this phase of the method (queries registering metaphysical concerns are added when relevant).

1. Since plurality* is a common feature of experience, there must be a target class as an initial principle of organization.
2. Plurality as such is suffused with differentiation*; we must therefore have some sense of what we are looking for in order to initiate collection with the expectation of an informative outcome. This factor is relevant since there must be some minimal limit on possible contexts when the methodology has reached a procedural point ready for division; if the collection were completely open-ended, there would be no parameters for guiding the requisite investigation of particular items to determine whether any formal samenesses are present and, if so, how they should be identified.

3. How many particulars must be collected in order to reach a "critical mass" for a legitimate collection? Will the precise number depend on the simplicity or complexity of the items collected (considered as individual items)? If we continue to collect (by addition, as in our sample Abelian group), when can collection terminate? When we do reach that juncture, the number of items collected will be methodologically sufficient to justify establishing a substantive metaphysical conclusion. The metaphysician must therefore determine a certain number of particulars in order to achieve "closure" on collection as the initial phase of the method. The items collected are denumerable, but the purpose of grouping is to investigate and determine the natures of the items gathered, not simply to count them. The number of items gathered may be relevant to the conclusions drawn after division has done its work; however, as such, as a purely mathematical dimension of the items under methodological scrutiny, the number of items remains independent from the number of properties derived from the items, given proper investigation.
4. Since the particulars exhibit differences among themselves, the respects in which they "associate" (or do not associate) in terms of relations with one another are crucial for success in applied methodology.
5. Particulars taken as such may be considered as units (the transliterated Greek for unit is *monad*, Leibniz's basic technical term). When particulars are initially collected, what will count as a "unit" measure?
6. If properties* of units can, in fact, be identified and properties differ between and among units, then these differences must be gathered into "some kind of class" in order to focus on what is common (the identity element) to all the units.
7. Will the standard that coordinates the initial collecting of particulars be based strictly on perception, that is, if things look similar, will they then be gathered into one collected group of items to be investigated? In general, is the unity* of a particular open to analytical description in multiple respects, for example, does it have discernible properties* insofar as the particular can be construed as a unit? These questions can be posed prior to collection as well as after the process of collection has

been concluded. (To claim that this unity is "abstract" does not clarify how the particular exists *as* a unit.)

8. *All* items in a group must be investigated (since an omission of even one item may skew the results of inquiry in terms of what is common to the items in a given group). Furthermore, *each* item in a group must be investigated in *all respects* (since each item in a group may be the same as all the others in every respect but one; thus, to omit this one disqualifying aspect—a methodological counterpart to the negative element in Abelian group theory—will affect the comprehensiveness of the final account of the formal structure proper to the items of a group taken in their full collective enumeration). If the initial collection is based on noticing some sort of similarity among items gathered (assuming the number of collected items is more than two—it may not be, depending on the goal of the method as applied in a given instance), then the range of apparently shared characteristics may be extensive. The process of winnowing out the relevant formal structures then begins and may be difficult to execute and complete properly.
9. Given the plurality* of particulars, establishing "sub-groups" of particulars will be useful to facilitate discerning the structure of smaller numbers of the group as parts of a larger, collective whole.
10. It must be assumed, presumably, that the units will remain the same over time; if their unity* is somehow affected by the passage of time, then the results of division may not be reliable in terms of ultimate metaphysical components.
11. Once the collection of particulars is completed, what kind of unity can be predicated of the class of gathered and settled units? Is the class equivalent to a Form of a certain specification, or is it some other type of hybrid entity (since its existence as a unity is temporary until the metaphysician completes the division of its content)? Is the unity of a group as a group affected by the identity of any one or more than one element in that group? If not, then a group can consist of any number of elements regardless how diverse they may appear to be when initially juxtaposed with one another. The number of elements

and the diversity emergent from the set of all elements, does not affect the feasibility that such a set with such members can be collected into a single group.

12. Will the quest for underlying samenesses—Forms—exhaust the gamut of reality of the collected items or should we expect to find among the totality* of particulars phases or aspects or properties of collected items that are "left over" yet which undeniably belong to the complete metaphysical reality of each item collected (in addition to the elements collected which, after division, yield Forms)? If so, is such metaphysical "residue" a problem for the explanatory integrity* of a given metaphysical system? (The diversification* of a system would seem to be affected in some respect if the interplay of Forms and particulars did not encompass everything that existed.) And finally, is the "being" of particulars subject to collection? If not, then if inquiry into being is a legitimate metaphysical concern, it follows that the methodology of collection and division is structurally limited and incomplete as far as discerning perspectives on reality which, at some point, will command the attention of at least some metaphysicians. The ultimate object of collection and division must have determinate content in some respect. It is worth noting then that thinkers may not be content with the principle that "being can be said in many ways" (e. g., Martin Heidegger, for whom the "being question," *Zur Seinsfrage*, is one of, if not the, primary concerns of properly fundamental philosophical activity—*Denken* in Heidegger's technical vocabulary).

Upon review, our enumerated list of factors involved in the collection phase of collection and division can be readily recognized as involved and intricate in matters of execution. We should not be surprised by these often subtle complexities; after all, the method of collection and division is intended to produce results that adequately reflect the structure of reality, even when this structure, once established, is superimposed over a world of vast numbers of divergent particulars. And, of course, the methodology under scrutiny is only partway done.

Division

The class of collected particulars is at hand. Now division begins, resulting in identifying Forms that constitute these particulars. To help clarify this concluding phase of the method, that is, when its application has produced results with apparent metaphysical authority, we discuss division by introducing descriptive headings for each step. Discussion of the steps is arranged according to an ordered hierarchy of themes based on available descriptions in the relevant passages from the dialogues as well as traditionally standard approaches to the metaphysical enterprise. Once we have reached this juncture in the methodology's application, the working assumption is that if the process of division succeeds, the Forms identified will be canonical, that is, they will exhibit properties of Forms as these properties have been described in other dialogues. The thematic arrangement of the following outline is intended to facilitate recognizing the degree of penetration achievable when employing a schema of evaluation aimed at metaphysical inquiry and derived from Aristotelian principles and concepts, elements which have also provided the conceptual means for evaluating aspects of Platonic philosophy of mathematics.

Unity and the Extent of the Forms

In Criticism A, Aristotle notes that the unity of 2 (or any number) is the sought-after element; the parallel goal in collection and division would be identifying the unity, stated at the level of Forms, which characterizes all the items gathered in a given class.

However, the unity of Forms becomes especially problematic when Form-numbers ground the existence of numbers. Since Form-numbers are constituted by units, the status and structure of units are crucial. For example, Aristotle investigates the number 10 (Argument A, Phase 3), which, as a singular entity, is comprised of ten units. As illustrated in Criticism A, his strategy is to question whether 10 could originate from two groups, each containing 5 elements. Such sub-grouping contributes to determining the cogency of explanatory elements advanced to account for the structure of Form-numbers. And as we have seen, Aristotle unearthed a variety of serious difficulties in this way of accounting for the existence of a given number.

Our current interest is whether, and if so the extent to which, such criticisms bear on the operations of collection and division. We note, for example, that the prevalence of appeals to unity as a factor in this kind of critical strategy must be considered with care. If, for example, a sub-grouping of a collection has a characteristic proper only to those items so collected, this characteristic may not be relevant to the structure of the collection as a whole with respect to its goal of identifying foundational Forms. It seems necessary then to distinguish between *partial* collection sameness and *total* collection sameness. The products of this distinction may not always be relevant to attempts to criticize an argument for explaining the origin of a given number. However, deployment of items into sub-groups in the context of division, especially when the items under scrutiny are more diverse in their singularity than numbers, may be directly relevant to identifying common properties suitable for formal designation. The use of sub-grouping as a strategic tool nonetheless facilitates recognition of such differences, should they exist. And, more generally, the Abelian factor of *associativity*—$(a + b) + c = a + (b + c)$—becomes useful as a formalizing and heuristic gambit for concentrating attention on a *limited collection of elements*, thereby facilitating the discovery of any substantive differences emerging from the elements connected according to this specification.

Also, in a related manner but moving in the opposite direction (from whole to part), we would have to assume that any sub-group of entities, as a self-contained unity within a more inclusive unity (e.g., a sub-group of 5 within a group of 10), will not affect the character of any of the individual entities nor confer on any one (or more than one) of the members of the sub-group some property solely by virtue of its (or their) membership in this subset. If this kind of "grouping" alteration would in fact eventuate for *some* member (i.e., in the logical sense, at least one), it could affect any claim of formal structure based on consideration of *all* the members of the original grouping of collected items. Thus, the members of a duo might be related to one another in a way (or ways) that affects how this one set of particulars interacts with other duos, or with other groupings of particulars with a larger number of items.

Now consider Forms as such, not sub-groups as a means of understanding the structure of Forms which are Form-numbers (Aristotle's immediate target in Criticism A). If a Form has a distinctive kind of unity, then determining, if possible, how many realities have this kind of unity becomes crucial. This is the standard problem of the extent of the Forms but couched in terms of the fact that a Form is a unity. As Aristotle points out (Criticism A, Phase 3), if there is a Form for the number 10, it seems there should be a Form for the number 11. And so on.

In general, how do we determine the range of relevant Forms insofar as such a range is necessary in order that division may account for the full reality of a given group of collected entities? But moving in the opposite direction of analysis, should we assume as a procedural priority that the ultimate goal of division is to isolate a *single Form* (replicating the strategy of Criticism A concerning a single number—10—as the access point of critical analysis)? Is it possible to clarify whether specification of a single Form is sufficient to account for everything that must be explained regarding all determinable characteristics of all particulars constituting the collected class* of particulars?

Whether division produces only one Form or several Forms appears to ignore the importance of the *Philebus* mention of collection and division (16d-e), which suggests that multiple Forms may be present in certain instances of collected uniformity. Also, Aristotle's critical treatment of the Platonic position is effectively geared for the multiple Form possibility; as a result, that will be the direction of our discussion. How then should the metaphysician, if considering the use of the method of collection and division, determine as a matter of methodological procedure the range of Forms necessary to account, by means of division, for the complete reality of a given group of particulars?

Questions concerning the extent and number of Forms with respect to division lead to the theoretical relevance of a series of issues raised in Criticism E. The various strategic moves in vocabulary and reasoning Aristotle introduces here exert pressure on determining the underlying metaphysical structure of a group of collected particulars, that is, distinguishing between (a) what would be common, existing at a level due to the causal agency of a Form, and (b) what "belongs" to units (as members of the collected group) but is not due to a Form. As it happens, Aristotle's axiomatic contention that one is a measure, not a number (1088^{a}3–6), has implications that throw into question certain aspects of the notion of unity, or oneness*, and these aspects are relevant in various ways for the functional activities of division in this regard.

Consider, for example, the plausibility of maintaining that the elements in one of Aristotle's examples in Criticism E–man, white, walking–can indeed be counted as three. Aristotle's reason for denying this possibility is that white and walking, as instances of categories, necessarily belong to man and man, as such, exists as one substance. A review of Aristotle's argument indicates that two different approaches are taken toward the sets of elements considered–first, when each element, in its entirety, differs from the entirety of each of the other two elements; second, when one element is considered but from three different perspectives. The unity of a single individual of a certain type thus supersedes the unity of distinguishable aspects of that individual to the point where none of those aspects can be counted as "one" reality.

From the standpoint of division, an advocate of Platonic methodology would respond that this sense of unity is too restrictive. If, for example, the subject matter of collection concerns human beings, then it may become necessary to identify certain aspects of each human being collected for purposes of determining whether all human beings so gathered exhibit the same set of aspects, and then to investigate whether the shared aspects can be accounted for by formal considerations designated by applications of division. At the point of initial gathering, each item as a member of the collected set stands as an individual unity, to be sure; however, the process of division approaches that individual unity and subjects it, necessarily, to partitioning such that the constitutive character of the unity becomes visible with respect to fundamental structure (i.e., a Form or Forms).

It should be noted, however, regarding the complete analysis of Criticism E that Aristotle eventually moves in the direction of a kind of "divisional" metaphysics similar to canonical collection and division. Julia Annas observes (JA, 197):

> Aristotle should hold that we cannot count together items in different categories. Objects, qualities, relations, etc. do not form a group of things

> that all exist, in the same sense of "exist." So, since counting involves at least the ability to reidentify things as a being of a certain kind, they do not form a group of things that can be counted either.

Annas dryly adds that "this idea is hard to sustain." It comes as no surprise then that in the end of the criticism, as stated, "we find Aristotle sensibly conceding that we *can* [italics in text] count the items—there is some unit measurement that can be applied, if only 'categorially different item.'"

We note, of course, that the proponent of collection and division does not want just *to count* collected items; what the proponent of this methodology intends is *to analyze the formal structure* of the collected items. We must be able to collect individual items, units, so that when grouped, it will be possible to determine the identity element, stated at the level of Forms, which is common to the collected set of individuals. Given the emphatic metaphysical point of this methodological step, it is evident that every individual in the group can and must be counted, one by one, as a prerequisite for membership in the group (this procedurally uniform treatment is equivalent to the process of admission and closure in the formation of an Abelian group). Without such a preliminary and summary identification, it will not be possible to examine all the units both as individuals and as members of a group. If during this process it should eventuate that one or some, individuals are not sufficiently alike or the same as other individuals, then each of those individuals must be set aside as not relevant to the metaphysical goals of collection and division *as pursued in this instance of the methodology*. This consequence is salutary for effective methodological practice. But now it becomes evident that if we follow the consequences of Aristotle's argument in Criticism E, a sense of unity* applying to any one item, constrictively construed as one substance, has overridden the primary goal of this method: when this one item is examined in conjunction with other items in a collected group, the point of the examination is determining whether some type of reality being predicated of the particular should (or should not) become the object of "group scrutiny" in order to discover the structure of this and all other collected particulars as determined by Forms.

It is the case that whatever "unit measurement" we can secure will satisfy the Aristotelian requirement that counting must be based on recognition of some unit factor, a unity which encompasses a reality exhibited by all elements of a given set (three, in the example at hand). However, here again an emphasis on the factor of unity in order to justify the process of counting imposes a single dominant underlying feature which the method of collection and division would not consider necessary; indeed, reduction of the "categorically different" elements to inclusion under a single more inclusive reality would, from the standpoint of collection and division, be considered a potential impediment to accurate metaphysical analysis. For even if a more inclusive "unit measurement" could be produced as an Aristotelian ground for justified counting, each element so measured must continue to be recognized *as different from* all other such elements. When these differences are identified and described, they will contribute to recognizing Forms as they inhabit and define the items constituting the set of gathered elements.

Aristotle denies that one is a number. But if we assume that the use of "one" is to determine appropriate metaphysical referentiality, Aristotle's position reduces to a variant on the linguistic ploy of replacing "one" with "measure." However, the proponent of collection and division would insist that the notion of "one measure" is not redundant—this noun phrase points to a context of inquiry such that, upon investigation and proper division, the metaphysician could identify a relevant Form that measures, or constitutes, the set of units collected for purposes of metaphysical analysis and description.

It may be observed that Aristotle's rigidly spare sense of unity in this setting seems, at least initially, to run against a line of analysis introduced earlier in *Metaphysics* VII. As a "this," or subject, each item gathered is one item; however, what for Aristotle appear to be derivative considerations as far as the essence of that item is concerned become, within the functional scope of collection and division, characteristics that may define the structure of that item. It is worth noting, however, that Aristotle's reduction of differences within the metaphysical dominance of one all-embracing unity can nonetheless become a galvanizing factor in the practice of division. The potentially stultifying effect of containment within a single unity when, in point of metaphysical fact, formal realities exist and await discovery and identification demonstrates the necessity of looking at the items collected from a variety of perspectives—deep and determinative in the quest for animating Forms. Resistance to the apparent force of the Aristotelian criticism in this context may indirectly allow hidden realities to be eventually discovered and identified.

Aristotle's example embodies three distinct types of emergent unity—entitative unity, categorical unity, and relational unity (i.e., the dependency of properties on something that serves as a metaphysical ground for these properties). Does division reveal Forms applicable only to a collection of *individual* unities, or will it also identify unities at the level of Forms for *properties of unities* if these properties are construed as separable types of reality? For Aristotle, categories explain these kinds of realities; the question is what collection and division will do with these data. Division must isolate whatever is characterizing the set of particular units that is due to a Form; the question then becomes whether some property (or properties) of a particular may be due to something other than a Form. The Platonist must be prepared to distinguish between a commonality that can be properly identified as due to a Form and a commonality that can be more properly identified as due to the presence of what Aristotle identifies as a category.

Forms and Hierarchical Structure

When appropriately applied to the class of collected items, division produces Forms. But how are we to understand the results if division produces more than one Form? How should the constituting Forms be understood in relation to one another? Are all Forms co-equal, or should we expect to confront the need for some sort of hierarchical arrangement? If the latter, then it follows that distinct types of differentiation* are factors in the division phase of the method; also, the brief programmatic descriptions of collection and division in the relevant dialogues do not emphasize or even mention such contrasts.

An example of such arrangement is readily available. As a matter of principle, Platonists posit a plurality of units existing as Form-numbers in order to account for the existence of numbers. In Criticism B, however, Aristotle contends that these units, inasmuch as they are completely undifferentiated, cannot function in that capacity. The criticism contends that in order to be theoretically feasible, the units constituting the number 2, for example, must be metaphysically distinguished in some way. Aristotle's proposal is to identify the units as existing in potency relative to the possibility that they can be combined as an actuality; the result, a combination of two different kinds of reality, will generate the required metaphysical duality and in the process generate the origin of the unique number 2.

As a critical response to the position in question, Aristotle introduces a different metaphysical scheme and claims that it can provide explanatory material that is lacking entirely in what the Platonist asserts. Stand back from this response and consider what Aristotle has done. In general terms, what we have is a direct confrontation between divergent metaphysical positions. The procedural assumption is that juxtaposing one position with another (i.e., belonging to a critic) will not distort in any significant way the structure and conclusions of the target position. Can the possibility of such distortion be eliminated, or at least reduced, when evaluating a metaphysical position?

Aristotle's strategy in this context raises a question of considerable import. We wonder whether any attempt to confront the methodology of collection and division (or, indeed, *any* duly formulated position) will necessarily require an already established basis in some metaphysical position, even if that position is only inchoate or implicit. If so, we would presumably want the confronting position to be the *least inapposite in terms of its own structure*. What Aristotle has wrought in this case does indeed provide a basis for distinguishing elements in a collection and, in consequence, establishing the necessary conditions for drawing the desired conclusion. But the strategy, as it emerges in the context of the argument and without any preamble, evokes at least a mention of the Straw Man fallacy. Thus, the target position is readily dispatched by introducing significant—but very different—metaphysical elements into the discussion, then claiming that the mere introduction of these elements is sufficient to demonstrate that the original position cannot be maintained as cogent.

This possible procedural matter to one side, the relevant critical force generated by Aristotle's line of argument depends on juxtaposing two different general elements which, in relation to one another, exhibit a metaphysical hierarchy. Potency *depends on* act, a fundamental metaphysical difference that must be not only recognized but also deployed in the correct order. Thus, Phase 1 of Criticism A cites the need not only for difference between types of things but also that the difference incorporates a dependency relation. Criticism C also comes into play in this scenario, especially with respect to specifying identity conditions for the proposed explanatory elements.

The potency/act difference is hierarchical in this respect; thus, if a species/genus arrangement—also hierarchical—is proposed, there will be the consequent necessity of identifying a differentia to separate one species from another given that distinct species fall within the same genus. This requirement, if legitimate for the successful application of division, presupposes that if all the elements under scrutiny are Forms, then the Forms can be configured with respect to one another in a hierarchical manner if such a configuration is required by the nature of the Forms generated by division. It follows that the unity of some Forms may involve a factor of self-identity by virtue of which that Form is the Form that it is; however, the same unity also incorporates *a relation to* another Form with its own distinctive self-identity. Furthermore, this relation connects the two Forms by dependency considerations. The unity of the Form under scrutiny will therefore include an internal differentiation such that it can function in two distinct senses: first, with respect to the Form itself by establishing its unique identity, and second with respect to reflecting a certain definite kind of direction to another related reality.;

The notion of directionality, as it might be called, stands as an organizational prelude to other dimensions of a functioning division once the methodology has reached that level of analysis.

Forms and Process

It is obvious but still worth mentioning as an offshoot of Phase 2 of Criticism A that even a successful division into Forms does not resolve the problems connected to participation when Forms exist in causal relations to the realm of particularity*. One aspect of this situation can be made more precise by considering participation from the standpoint of unity. If a grouping of things is, for example, made a dyad because the things share in twoness, is the process of "sharing in" any different (that is, less or more problematic) than the participation relation for any kind of causal interaction between particulars and Forms? For example, is each element in a dyad a unity* in precisely the same sense as the unity of the dyad itself? Each element in a dyad retains its unity as such, as an individual element, but each element in a dyad also exists in a conjunctive relation to another element, that is, the other element in the dyad. Thus, unity can refer to the mode of existence of the element as such, as this element, but it can also refer to the mode of existence of an element insofar as it exists in conjunction with another element as joint members of a different, that is, dyadic unity (identical twins may illustrate this metaphysical medley). However, it would seem that the unity of a dyad would be unaffected *qua* unity if either unit in that dyad were to change in any respect or even to be considered as a member of a relation (which necessarily involves at least two elements). Whether this inference derived from the factor of unity usefully pertains to the question of participation is perhaps worth some reflection.

A second aspect of process assumes relevance at a point prior to the context of Forms as the object of division (although, as we shall see, the process factors to be discussed will directly affect what division will be in a position to identify). If, as we maintained in the previous section, the appeal to potency represents one way to establish a hierarchy among related Forms, should this hierarchy be introduced into the general structure of collection and division? For example, does the collection of individual units ignore the entire dimension of potentiality, of the units becoming something other than or more fully realized than the state of being of the item when it was actually and initially collected? If collection occurs "in the moment," that is, within a determinately brief period of time, does this factor of temporality affect an appeal to potency insofar as the period when some particulars are "in process" to become something which, at the time when collection occurred, they had not yet become? If so, then collection by itself becomes a more complex type of activity by virtue of the need to introduce a process dimension as an integral feature for capturing the complete reality of collected items.

A third scenario: Criticism A Phase 2 indicates that some things come into existence without any connection to Forms. But if certain existing particulars are not amenable to being analyzed by the application of collection and division, how does such entitative independence affect the integrity* of the system generated by applying this method to the totality* of particulars? It would seem incumbent on a philosopher seeking comprehensiveness*, especially with respect to systematic inquiry, to complement any such deficiencies in the effective range of collection and division by adding another method, or at least some sort of approach to those "outlying" realities so that they can be accounted for metaphysically. The lesson imparted from these considerations is that investigation must determine whether such methodological "gaps" in range of content do in fact characterize the application of collection and division. If so, what would constitute a responsible reaction to this situation?

Forms, Properties, Predication

A unit as a component of a Form-number is "abstract" (although we are uncertain how to take this ascription), but a unit as a particular entity in the world is replete with properties. Aristotle does not believe that a Form, as its own kind of unity, can endow a particular with all the properties that a particular, as a unity, possesses. In Criticism C, Aristotle announces the need for categories as avenues for rendering particulars adequately in terms of how each particular displays all the different aspects belonging to that particular as a unity, as a singular reality existing in concert with a myriad of other realities both like and unlike itself. Particulars instantiate a complex constitution of units with respect to their unity*, with the resultant possibility (for Aristotle, necessity!) that accounts of their comprehensive reality will require more than Forms as explanatory and descriptive vehicles.

How, Aristotle wonders, can one thing, even a richly defined reality such as a Platonic Form, cause so many different properties? If a Form is posited as accounting for the comprehensive* reality of a unit (and other units marked by a sameness common to all units in a given collection), does this Form also account for *all* distinguishable aspects of *each* unit in that collection? (Recall from the review offered above listing factors involved in collection that the particulars originally collected differ from one another in at least one respect.) If all the properties of a particular are conferred by one Form, then the oneness* of this kind of reality must be of prodigious metaphysical endowment (including its efficacy in the realization of a proliferation of diverse particulars with, taken collectively, virtually unlimited property diversification*). The range of raw material potentially falling under the domain of methodology *prior to* an application of collection is daunting in various relevant respects.

The reply might be that Aristotle is imposing too much for a Form to accomplish. The statements of collection and division in the dialogues stipulate that the Form will account for only what is *the same* in a group of particulars, not for *every single aspect* of reality exhibited by the collected items.

This clarification certainly seems to reduce the theoretical extent of the problem, but for Aristotle this ploy will doubtless not be enough to increase the cogency and persuasive power in the appeal to Forms. For now, the problem has become distinguishing between (a) Form functioning as a kind of foundational cause (of a given sameness in a set of collected particulars) compared to (b) available perspectives on a particular which are "part" of that particular's metaphysical profile but not in the same way that a Form is "in" that reality.

Consider the fact of coincidences. In the Aristotelian way of things, coincidences are common to different units, but it does not follow that a number (such as 4 as the number of the seasons) has anything to do with what that number is said of. For Aristotle (as argued in Criticism C), all supposed likenesses Platonists claim to associate with certain numbers "are incidental (though all appropriate to one another) but one by analogy. For there is an analogous item in each category of being." Appropriateness is present in such juxtapositions, but the "one" that the Platonists claim results from this degree of appropriateness and thus is somehow privileged is so only "by analogy." This is, in effect, a unity without portfolio as far as establishing itself as a metaphysically credible characteristic with significant explanatory heft. Hence, for Aristotle, the need to specify categories in order to do justice to the complete unity* of a particular.

A more workable explanation rests on the fact that there is "an analogous item in each category of being." Categories accomplish the substantial predicational work; all they do, however, is point to linkages between and among particulars which are characterized by similarity and analogy but not by conditions of sufficient sameness to warrant the introduction, the necessary introduction for the Platonist, of Forms to account for them. The problem may be stated in detail thus: when the units collected are examined from the standpoint of division, how can the method distinguish between (a) a sameness necessary to the existence and nature of the items collected and thus due to a Form and (b) a characteristic which may be equivalently common to all the items collected but is only coincidental, that is, it happens to be common to all the units but is not sufficiently substantial to warrant status as a Form performing at a fundamental level regarding the existence and nature of the items collected?

In essence, Aristotle insists that Forms have limited explanatory value depending on the context of inquiry. When confronted with this broadside of concerns, a prospective practitioner of the methodology of collection and division seems driven to embrace a kind of metaphysical miscellany regarding the totality* of particulars along with equivalently wide-reaching and divergent implications for comprehensiveness* as these factors pertain to achieving reliable explanatory results. Consider then, for a moment, a methodological possibility that integrates the Aristotelian option, to a point.

One approach offering at least a modicum of interpretive control over the seemingly limitless demands of basic experience is the ready employment of categories (as, indeed, Aristotle has insisted). We may view categories as ways to break into the unity of each collected item in order to discover whether a Form (or Forms, interacting in some way) is (or are) present. If this procedural possibility looks attractive, would we then need a pre-established categorical scheme in order to pursue collection and division, or could we rely on the inherent modes of usage embedded in ordinary types of discourse (e.g., nouns, verbs, modifiers)? If the results of collection and division are expressible in ordinary language, does it follow that ordinary language is sufficiently perspicuous to fulfill the goal of this method in order to produce a fully comprehensive formal account of each set of collected items? And a second, perhaps more penetrating and fundamental concern: how would division distinguish between a category as a register of a condition or characteristic, which Aristotle identified as something incidental, applicable to a group of particulars (and capable of being expressed in language), and a Form that functioned in more fundamental respects?

The differentiation* of particulars becomes apparently even more convoluted if we concentrate on the place of language and discourse in both life and philosophy and draw the important distinction between the unity of a particular with respect to its metaphysical origin versus the same unity with respect to predication (i.e., "how we talk about things"). If we invoke the sense/reference distinction and introduce changes of usage depending on culture and history, then predication would seem to be a phenomenon that was and is, in some important senses, continually changing. Questions regarding predication are independent of questions regarding causality; how the unity of any particular is understood will have an extensive effect on the theorizing offered to account for these two related but distinct metaphysical matters. In general, a different kind of engagement with a different kind of reality characterizes predication insofar as it denotes a set of circumstances involving the vagaries of language as well as tight connections to general terms; these circumstances are distinct from the ways particulars exist in relation to Forms. But, again, predication—especially in the contemporary philosophical world—must be among the major areas of investigation as far as overall systematic credibility is concerned.

It seems then that Aristotle's line of criticism in this context should be acknowledged as useful in terms of pointing out structural possibilities—indeed, perhaps inevitabilities—in the application of collection and division, as well as, perhaps, indicating implications that may bear on inherent limitations of this methodology.

Forms and Relationality

In the *Phaedo*, Socrates makes a point (noted in Chapter 1) of the fact that the Form of threeness not only causes the number 3 to be what it is, but it also confers oddness on that number (104d). Therefore, when in Criticism D Aristotle refers to the fact that numbers are involved in the relations of "before and after," evidence exists showing that Plato was aware of this dimension in the reality of numbers and thereby asserted that a Form can confer a basic unity (threeness) as well as a property (oddness) to the same number.

The question becomes whether this dual directionality is sufficient to account for both "before and after" as well as, say, "greater than" and "less than." These are different kinds of differentiation* illustrating two distinct types of relations regarding numbers. And there are, of course, many kinds of relationships, some mathematical, many non-mathematical.

Consider "is greater than." Relations involve plurality* In order to account for relationality*, do we want Forms dependent on something other than themselves? If not, then perhaps we might consider relationality conceived in the Leibnizian mode as a property of an entity. Thus, the property "1 less than 4" belongs to 3 as "part" of the complete reality of the number 3 (as does the property "1 more than 2"). The > and < relationships could perhaps also be so interpreted. Would we then generalize this circumstance and contend that Forms do not merely cause the existence of particulars (as the *Phaedo* asserts) but that any relational properties pertaining to particulars are also grounded in an appropriate Form? Thus, would, for example, being a parent, mother, father be considered aspects of items existing "in" an item in such a way that the predication of the relevant relation presupposes the presence of a Form? If so, then any division which produced a Form involving a relation of any sort would seem to entail that all such Forms exist in conjunction with something other than themselves as a necessary condition for generating the relational aspects proper to certain particulars.

As a relevant aside (and ignoring in this context non-Abelian groups), we observe that the Abelian group factor of commutativity ($a + b = b + a$) has interesting consequences in this context. If relational properties belong to given items in a collection, commutativity implies that reversing the order of two such items would *not* affect their "relationality." But of course, it not only affects relationality; relationality of this sort would in fact be destroyed. Commutativity is nonetheless useful from a heuristic perspective. The juxtaposition of two such items would allow an immediate realization that there was a relation present and functioning within the items in that collection; as a result, the collection could not be reduced to a simple enumeration of the constituent elements, that is, counting them individually as units. The elements had to be analyzed as units, of course, but also as units with additional specifications in terms of how their unity existed. (It may be noted that to put commutativity into practice in this fashion presupposes a metaphysician's awareness of some specific features of reality, which may be a factor militating against its heuristic possibilities.)

The line of questioning concerning Forms with respect to relations can be readily extended. In Book N (1090^b19–20), Aristotle is pursuing contexts involving numbers and relationality, and he observes (a passage discussed from a different perspective in Chapter 5): "However, judging by what we can observe, nature is not a mere series of episodes, like a bad tragedy." As stated, this oblique criticism exhibits a puckish charm in comparing the lack of internal organization of an inferior literary work to the impressively organic regularity displayed by the natural order; the relevance of the image to the issues at hand concerns how different elements or items relate to one another and to their membership in a more inclusively populated arena of collection.

In Phase 3 of Criticism A, Aristotle suggests that a sequence (e.g., the decade, the numbers in order from 1 to 10) be considered a reality by itself. In this reality, each number exists not only in itself, as a unique entity, but also concurrently as a member of a relation moving in two directions—is greater than and less than. In order to preserve this sequence, each item must exhibit not only *inherent* properties but also *relational* properties with respect to items other than itself. Thus, 2 as one number in this sequence is concurrently greater than 1 *and* less than 3, and so on. One reality exists not only as such but also in a pair of distinct types of relationality* to realities other than itself.

Aristotle also wonders whether in this regard the sequence, as a unity of individually different realities, is apparently more real (is "more of a real thing") than, say, the number 10 appearing by itself (although 10 is, necessarily, the final number in this sequence of numbers). The sequence exhibits a kind of sequential unity, locking in a certain number of units so that they follow a given relation and series of relations for certain definite intervals; as such, the sequence determines the appearance of each individual unit as well as the location of that unit with respect to all other units comprising this sequence. A sequence therefore exhibits controlling force, a property* which can be viewed in its inclusivity (per Aristotle's suggestion) as more dominant than any single member (or groups of members) within that sequence as a totality.

The sequence from 1 to 10 is composed of numbers and, as such, the structure and relationality of its elements may be readily recognized. However, Aristotle's approach to a sequence as a unity unto itself raises related questions in terms of collection of particulars and the subsequent interest in dividing this collection according to Forms. Thus, we may not appreciate the full extent of the formal presence characterizing a given set of multiple items *upon the initial collection of those items.* Only when concentrated reflection allows all the members of the set to begin to intermingle with one another—an integral phase of the process—would the metaphysician be in a position to discern a commonality, or commonalities, which may not have been evident at first reckoning.

An expansion of Aristotle's comparative literary image will illustrate the point. Initially, a tragedy may appear disjointed and lacking thematic or plot unity, but then, upon additional review and reflection, the viewer (or, indeed, the student) comes to recognize that what seemed episodic and disconnected may well be coherent, unified, and organically revelatory. In the same way, the items in a collected set may appear relatable to one another only in some minimal sense (without such surface interaction, as it were, the items would not have been gathered for methodological purposes in the first place), but then additional scrutiny may reveal a deeper, more inclusive slate of formal realities pervading the collection of particulars, an important instance of what we here have called relationality.

If these circumstances are accepted as relevant to the procedural dimension of collection and division, then it seems incumbent on the advocate of this methodology to consider including *time factors* in the execution of the process of collection. The length of such procedural dimensions of temporality will vary depending on the kinds of particulars collected and also, perhaps, on the specific goal of given applications of the method in terms of the desired Forms to be identified. But the heterodox quality of particularity* along with the properties individual particulars display is a factor which, by itself, would justify the methodological and practical relevance of time in certain, perhaps many, applications of collection and division.

This is the second aspect of temporality our investigations have disclosed when examining collection and division as operational procedures. The first aspect arose in the analysis of Forms and hierarchical structure when we noted that the realization of something potential with respect to the unity of a particular will typically require some sort of duration and in that regard could be relevant to the specific character of what will be gathered by collection. The two aspects of time differ in that actualization of potentiality refers to the units subject to collection (the "objective" side), whereas a period of concentrated review before a dimension of unity can be discerned pertains to the cognitive activity of the metaphysician at work applying the method (the "subjective" side). Aristotle's analytical appreciation for the distinctive reality of sequences may be credited as a stimulus leading to the recognition of this factor.

A brief recapitulation: Relationality has a double influence on division and the identification of Forms. First, the identification of a Form involved in some manner with this reality presupposes the existence of something other than the Form itself; second, this something other (presumably another Form, although this limiting specification should be examined with care) must be related to the given Form in certain ways. Could the relation itself be a Form, or is the statement of a relation simply an event in language without generating metaphysical implications of this foundational sort? Regardless of everything eligible to be analyzed in order to deal with that question with even moderate competence, relationality presupposes plurality, and it may also presuppose one Form interacting with another Form. This situation may be taken as a miniaturized and foreshortened setting for what follows.

Mathematics and Unity—The Threshold of the Good

At the conclusion of the previous chapter, we shaped a metaphysical mosaic, a configuration of terms employed by Aristotle in service to his critical evaluation of Platonic philosophy of mathematics. This multiplicity of terms represents a rich and diverse vocabulary which, as noted in commentary on the mosaic, commonly appears in many treatments of reality, whether approached as a totality or as sectors or perspectives on totality. Such ventures, grand and limited, have characterized the history of Western metaphysics.

Another summary approach to the metaphysical enterprise moves in the opposite referential direction—toward a singularity rather than a multiplicity of terms. A primary instance of this approach is the term one, or in a more substantial sense, unity. W. D. Ross, discussing possible senses of unity in a passage in Book N, observes (Vol. II, 475): "The unity may be (1) a unity of substance, (2) a unity of quality, &c., separately from substance, (3) a unity of all the categories together." Julia Annas provides another example of the variability of unity (JA, 145): "Aristotle has not distinguished two questions which are in fact rather assimilated by the Greek, 'is this thing one (*hen*)?' This can mean, 'Is this thing unitary?' or it can mean, 'Is it a unit (one in number)?'" For Annas, the argument under scrutiny at this point "runs together both these types of consideration." Lack of precision in determining the relevant sense of this term will presumably compromise whatever conclusion that argument ostensibly maintains.

The appeals to unity from Ross and Annas represent only the beginning in determining the boundaries of this area—or, perhaps, arena—of metaphysical discourse. The following is a set of concrete, seminal instances illustrating the fundamental Aristotelian principle that unity can be said in many ways, especially in contexts intersecting with mathematical entities construed as singletons and gathered in groups, and also as featured commonly in the arguments discussed in Chapter 6.

1. Unity of a Form;
2. Unity of a group defined as Abelian;
3. Unity of a single element of an Abelian group;
4. Unity of an item in a Platonic collection;
5. Unity of the class of collected items, from which class Forms are divided;
6. Unity of any single item referred to in the metaphysical mosaic (with the obvious rider that this list hardly exhausts possible metaphysical determinations at an equivalent level of generality);
7. Unity of predication, when any one property or class is said of any one particular;
8. Unity of a single theoretical complexity (e.g., group theory, a theory of predication);
9. The unity of the good as a foundational factor in metaphysics (is the good static in all respects, or is it capable of variation or change of some sort?);
10. Also, perhaps, the conjunction of unity and the good.

In a strictly Platonic context, the unity of a Form is one of its cardinal properties; it also stands as the theoretical locus of a paradigmatic problem given that one Form causes many particulars to possess instances of that Form—how then does a Form maintain its unity, its uniformity, its oneness?

A group is another kind of unity, also generating its own panoply of problems: thus, is the unity of a group distinct from the unity of any one element in the group or the conjoint unity of any two elements (or more than two, depending on the total number of elements in the group)? A related question: is the unity of a group as a single source of gathering affected by the identity of any one (or more than one) element in that group? If not, then a group can consist of any number of elements regardless how diverse they may appear to be when juxtaposed with one another. The number of elements, the collective diversity emergent from the set of all elements, does not affect the feasibility that this kind of set can be gathered into a single group. Furthermore, a group may appear to contain elements more like than unlike one another; however, despite this initial span of sameness, scrutiny may indicate that one or more than one of the originally gathered items is sufficiently unlike the other items to be excluded from the group. From this perspective, the unity of a group is not merely for the sake of sheer enumeration but to underlie the possibility of discovering resident commonalities to ground uniform descriptions at the level of language (if not an account of existing types or classes at the level of reality). Sameness and difference are thus crucial variables with respect to membership in a group and the utility of a group for metaphysical purposes.

The imposition of unity, or "oneness," as a kind of limit on difference and plurality appears to focus aspects of analytical scrutiny on the members or elements of that plurality, both in themselves and collectively, as they interact with one another—or, indeed, resist such types of activity. Is it possible then to advance a programmatic approach to unity and its cardinal relevance to the metaphysical enterprise that will orient its function and facilitate its use at high levels of speculative generality?

As such, the notion of unity, especially in the contexts of the range of realities discussed in this book, encompasses a luxuriously stocked array of possible developments. If so, then identifying any element in the metaphysical mosaic as a unity pinpoints that element so that it becomes necessary to situate it within process dimensions just to determine what the limits of unity will prescribe for that element, that is, how much leeway would be allowed to each element without losing whatever warrants the assignment of its unity.

These brief concluding reflections on distinct types of unity have culminated in a set of descriptors listed above and (a) derived from the various originating instances of unity (basic elements in arguments examined in this chapter) but also (b) ascending to higher levels of generality. This mode of progression, if allowed to continue in this direction, is moving toward some sort of foundational reality. Those metaphysicians with philosophical sensibilities primarily, if not exclusively earth-bound and rooted in concrete particularity will hesitate to embrace this style of theorizing. That proviso in hand, the reader is asked to advance to Chapter 7 with an open mind toward the possibility that a foundational approach to mathematics, if developed in certain directions, will establish possibilities for pursuing and perhaps realizing a more comprehensive and penetrating metaphysical vision.

7

Mathematics and the Good

A Speculative Prelude to Metaphysics

Chapters 5 and 6 each concluded in summary fashion with distinct collections of particularized aspects of metaphysics. These aspects assume increasing relevance if we view metaphysics as a sub-discipline of philosophy with its own history, from its origins with the Greeks into the present. The specifications—a metaphysical mosaic comprised of a wide variety of foundational terms (Chapter 5) and a series of perspectives on unity deployed within a wide variety of contexts (Chapter 6)—constitute a broadly articulated horizon of metaphysical elements, principles, reasoning, arguments, and critical conclusions. These materials were engendered by sustained reflection on basic features of the philosophy of mathematics found in Plato, Aristotle, and Whitehead in juxtaposition with the modern mathematical positing and development of group theory. The question may then be posed: Can this horizon of mathematical significance for metaphysical inquiry be extended even beyond the limits established by implications drawn from diversifications proper to Abelian groups?

We address this question after reflection on a core reality for this, or indeed any, aspect of metaphysics, an inquiry that becomes compelling especially if due weight is given to the history of metaphysics. To this end, the types of unity laid out at the conclusion of Chapter 6 are arranged to represent a gradual increase in referential range, thereby indicating a movement toward a base of metaphysical inquiry. Chapter 7 will demonstrate that a reconfigured, recast account of the good (modeled in part by the *Republic*'s evocation of this reality) will provide foundational elements showing senses that, and the extent to which, mathematical elements when appropriately modified can apply to metaphysical positions.

The Relevance of the Good

An examination of the influence of the good in matters involving a supposed confluence of metaphysics and mathematics may begin with proper acknowledgment of Plato. The good [τό ἀγαθόν] is the most basic and comprehensive element in Platonic metaphysics, with implications encompassing the Forms as well as everything existing caused by or predicated of the Forms. The subjects of mathematical inquiry also fall under this broad schematic, as we shall see in a review of the divided line in the *Republic*. In addition, the good identifies an element Plato, Aristotle, and Whitehead refer to explicitly in mathematical settings. Thus, all three philosophers hold that the good is an indispensable factor in understanding the foundational reality of mathematics. Such shared appeals should nonetheless be approached with care. It would be premature to assume that diverse philosophers all mentioning "the good" would be sufficient textual evidence to conclude that this trio had precisely the same sense of the good in mind when they entertained its importance with respect to mathematics. That reservation in hand, it nonetheless seems fair to say that *any* juxtaposition of mathematics and the good appears to be an indiscriminate mix of disparate realities. However, given the epochal authority of the thinkers who asserted precisely such a juxtaposition, this conjunction surely deserves a certain measure of sustained attention.

We therefore initiate this phase of the argument by reviewing seminal positions in the treatment of the good by Aristotle, Whitehead, and Plato. The array of texts will be considered in thematic order (rather than according to historical chronology) in order to highlight an organically apposite approach to the question of whether, and how, to extend the critical analysis of metaphysical phenomena through a series of general and abstract elements that also underlie mathematical realities. The position to be argued is inspired by identifying features common to both disciplines and referring to structural considerations assignable to a speculative account of the good. Since the good is a frequently employed primary element in traditional metaphysics, we may anticipate that these common features will be expressed in terminology pitched at a high level of generality and, furthermore, that they can be recognized to exhibit parallel functions in mathematics as well as metaphysics.[1]

Aristotle

Toward the conclusion of *Metaphysics* Book N (1091^{a}30-32), Aristotle refers to a "puzzle" concerning how the elements and principles of mathematics relate to the good and the beautiful;the relevant passage identifies an element Aristotle explicitly refers to in a mathematical setting (see his motto for this essay). It is, in a way, puzzling that Aristotle saw fit to mention this approach to the philosophy of mathematics at all, given that his own stock of preferred metaphysical elements in this area of inquiry remains closely aligned with the immediacy of concrete particularity, a tendency working in tandem with the rather severe limits Aristotle placed on the explanatory relevance of general terms (exhibited in some detail in the discussions of Chapters 5 and 6). One might nonetheless wonder whether shadows of some sort of transcendent good reemerged at this juncture of *Metaphysics* N, however tenuous their possible effect on the systemic treatment of mathematics undertaken in this work as a whole. The apparent incongruity becomes all the more intriguing if Aristotle intends, even if only indirectly or in a marginal sense, to evoke the Platonic notion of the good as it appears so prominently in, for example, *Republic* VI. The import of this question may be compounded when we observe that Aristotle raises similar possibilities much earlier, toward the beginning of Book M. Thus "those who assert that the mathematical sciences say nothing of the beautiful or the good are in error. For these sciences say and prove a great deal about them . . . . The chief forms of beauty are order and symmetry and definiteness, which the mathematical sciences demonstrate in a special degree" (1078^{a}32-^{b}2). This approach to beauty in a mathematical context is also of interest; indeed, the motto's spare introduction of "the beautiful" (along with the good) as the beautiful pertains to mathematics introduces a theme reappearing later, in Chapter 8. But the isolated, perhaps unexpected Aristotelian reference to the good may be taken to inaugurate the most speculative phase of this book.

How does the Aristotelian appeal to the good and the beautiful play any role concerning the overall goals of this book? We will pursue this question in Chapter 8, after establishing additional foundational elements so that an Aristotelian contribution in this regard can be more adequately appreciated. For now, the relevant point is just recognizing that Aristotle introduces a connection between beauty and the good at the beginning of Book M as well as toward the conclusion of Book N, thereby bookending the breadth and detail of his critical responses to Platonic philosophy of mathematics with these two broadly construable realities. The fact that Aristotle did not at either juncture in the *Metaphysics* analyze and articulate the asserted connections between beauty itself and mathematics can hardly be held against the range of interests he displayed in the general area of philosophy and mathematics; indeed, that he took time to register these connections at all is surely worth our attention.

Whitehead

In 1941, Alfred North Whitehead contributed an essay to the volume devoted to him in *The Library of Living Philosophers* (his health precluded replying to each of the philosophers who had essays in this volume). The essay, fifteen pages in length and divided into thirteen sections, is titled "Mathematics and The Good."[2] Here is the opening paragraph:

> About two thousand three hundred years ago a famous lecture was delivered. The audience was distinguished: among others it included Aristotle and Xenophon. The topic of the lecture was The Notion of The Good. The lecturer was competent: he was Plato.(666)

As a matter of principle, Whitehead asserts that "the peculiar associations of mathematics and The Good remains an undeveloped topic, since its first introduction by Plato" (666). Whitehead proposes to develop that topic through the notion of "pattern." Insofar as a pattern is, generally speaking, "known to finite discrimination," Whitehead's particular notion of pattern in this book refers to "a partial disclosure with an essential relevance to the background of the Universe." A more specific link between mathematics and the good is derived from the way in which pattern intersects with value: "All value is the gift of finitude which is the necessary condition for activity. Also, activity means the origination of patterns of assemblage, and mathematics is the study of pattern. Here we find the essential clue which relates mathematics to the study of the good" (674). Mathematics studies patterns and patterns reveal how elements in the universe are assembled; thus, mathematics is directly associated with the basic components that constitute the interactions characterizing the complexity and comprehensiveness of the universe.

Group theory is not explicitly mentioned in this essay, but Whitehead does analyze algebra which, as a branch of mathematics, plays a special role with respect to patterns. In algebra, we can write "x + y = y + x" (i.e., commutativity, one of the defining features of Abelian group theory) where "x and y are any two numbers." Thus, the emphasis on pattern, as distinct from the special entities involved in the pattern, is increased. Thus, algebra in its initiation involved an immense advance in the study of pattern" (676). The elements x and y are general; as such, the variability they represent stands as a kind of universality with seemingly unlimited instantiations. We see then how patterns, via algebra, extend their import into an abstract realm.

Whitehead draws an equivalently broad conclusion from this aspect of algebra and its relevance to pattern: "Thus the infusion of pattern into natural occurrences, the stability of such patterns, and the modification of such patterns are the necessary conditions for the realization of the Good" (677). The historical importance of algebra in the emergence of ordered patterns in mathematics only introduces the importance of "stability" and "modification" of patterns in other branches of mathematics (as well as science). This passage suggests that "the realization of the Good" with respect to mathematics will involve extensions from algebra (and group theory) to other branches of mathematics. Indeed, Whitehead asserts in a generalizing vein that "mathematics is now being transformed into the intellectual analysis of types of pattern" (677). We will pursue this possible extension of the good in mathematical settings, that is, what Whitehead refers to here as "types of pattern," later in Chapter 7 and also in Chapter 8.[3]

Whitehead summarizes his position on the good and the place of mathematics in the realization of the good as far as the dominance of patterns is concerned:

> Mathematics is the most powerful technique for the understanding of patterns, and for the analysis of the relationships of patterns. Here we reach the fundamental justification for the topic of Plato's lecture. Having regard to the immensity of its subject-matter mathematics, even modern mathematics, is a science in its babyhood. (678)

Immediately after this passage, Whitehead details the importance of what he calls mathematical understanding: "If civilization continues to advance, in the next two thousand years the overwhelming novelty in human thought will be the dominance of mathematical understanding" (678). If mathematics was in "its babyhood" when Whitehead wrote this book (not long before his death), then mathematics may well still be at that point in its history to this day. That said, mathematics continues to "grow up," as it might be put, and the fact about mathematics that will be explored below is the metaphysical condition for such a disciplinary-wide process of ongoing maturity, that is, that mathematics has its own history. More will be said on this theme later in Chapter 7.

Plato

Whitehead's thoughts on the seminal connection between the good and patterns, if viewed from the perspective of the history of philosophy, assumed that Plato's lecture on the good did indeed occur. Whitehead also assumed that the reports were reliable concerning the content of this lecture, that is, that it dealt with mathematics in a context in which the good was instrumental and integral. For the purposes of this book, it is important to note the existence of this lecture, but it is also important to state its relevance to the essay's basic argument.

Julia Annas offers the following on this theme: "The importance of the good as a basic principle for Plato is apparent in the emphasis in the *Republic* on the Form of the Good." She also reports that a variety of scholars claim to have found "in the fragments of *On the Good*, a theory of rational and ethical structure which underlies all of Plato's written works." However, Annas insists that "we know very little of the exact role of the good" in this lecture. As far as firsthand evidence is concerned, we know "from a report of Aristoxenus, who got it from Aristotle, that the audience was finally baffled by 'the limit, that good is one.' This phrase, however, has been translated and interpreted in many ways, as identifying the good and one, as saying that goodness is unitary, that unity is good, and even that limit is one and so good" (JA, 214).

In this book, we discuss the "Form of the Good" later in this chapter. We will also introduce a number of ways to approach the concept of one or unity with respect to the intersection of mathematics and metaphysics. Group theory, the dimension of mathematics introduced in Chapter 3 and employed throughout Chapters 5 and 6, will provide a foundation for more extensive applications of mathematics all under the notion of the good to be formulated later in this chapter. *On the Good* was a lecture, not a dialogue. However, its purported basic themes—the good, mathematics, unity, the configurations of patterns—are foundational for this book. These realities, occurring when they did in Plato's life, also provide a kind of continuity of theme from the Plato of the middle and late dialogues. It should be noted now that the theoretical, speculative adaptations of the good that will drive the argument in Chapters 7 and 8 are, in several ways, very different from the canonical treatment of these themes as appearing in the dialogues. But the lecture on the good provides at least *prima facie* evidence that these basic realities remained on Plato's mind even after decades of writing in dialogue form, thereby pursuing through this concerted labor measures of clarity and cohesiveness regarding the intersection of these realities.

Summary: Aristotle announces the importance of the question regarding mathematics and the senses, even if in the end it is tenuous, in which mathematics is related to the good; Whitehead celebrates the crucial character of the good in mathematical contexts, epitomizing its significance by appealing to the essential function of patterns in all mathematical thinking; Plato initiated this line of thought by, it is believed, verbally engaging an audience interested in the good and expounding—presumably in some detail—how the good and mathematics intimately interpenetrate one another. This common theme—the relation between the good and mathematics—therefore commands attention.

In this chapter, we investigate the possibility of generating a framework that will incorporate these concerns on a more inclusively determinate scale. The relevant implications will point to, and justify, an expanded universe of metaphysical inquiry based on a mathematical foundation.

One key question emerging from the preceding collocation of positions, especially given their high level of generality, concerns how to articulate the good so that its central effects will direct inquiry into the relation between the distinctive nature of mathematics and attempts to disclose the broader, more inclusive structures of reality as such. This question leads directly into the discussion of the good in the next section.

Perspectives on the Functionality of the Good

The perspectives in question refer to texts in the *Phaedo* and *Republic*. We will move from discussions concerning the good in these dialogues toward a reformulation of the good (and, presently, its renaming) informed by the goals of this book. The following is not an attempt at Platonic exegesis; rather, it represents derivations from Platonic texts that can stand by themselves as a speculative attempt to situate and broaden the argument of the essay. The *Phaedo* sketches an intriguingly global perspective on the good; the *Republic's* treatment of the good is more developed and also has the advantage of juxtaposing in close proximity the domains of mathematics and metaphysics. As a result, the following remarks are limited to speculative implications drawn from these textual givens.[4]

The Good Holds All Things Together

In the *Phaedo*, Socrates delivers a wide-ranging account of causality. He also criticizes competing attempts to explain causality, and the most fundamental objection is that these accounts "give no thought to the good, which must embrace and hold together all things" (99c6). Socrates adds that he would be more than willing to be a pupil to someone who could teach him a cause in this fundamental sense, but he failed to learn it himself and also could not locate anyone who had this level of knowledge. He does offer to provide an account which he calls his "second sailing" in quest of this cause. The remainder of the *Phaedo* follows this statement of his proposed agenda.

Although brief, the *Phaedo* passage conveys an important function of the good. The basic strategy of the book is to move in the direction of a perspective on reality which, when duly qualified, will sanction a more inclusive approach to appreciating the relevance of mathematics to the metaphysical enterprise. Attaining this speculative goal entails a certain interpretation of the "all" of "all things" [ἅπαντα]. Thus "all" could refer to the physical world by itself, to the relation between the "two worlds" (recalling the standard description of the metaphysics of the *Phaedo*), or to the Forms as such, independently of instantiation and also considered in relation to one another *qua* Forms (which would also entail difference as a necessary feature of this kind of discrimination). If the scope of the good were exhaustive, the good would also hold together the Form-numbers as separately existing realities.[5]

We shall take the *Phaedo*'s spare but in its own way spectacular claim as if the intended scope of the good functioned in all these possibilities and at all the levels of reality just enumerated. We recall the *Timaeus*' account of natural things characterized by mathematical elements at an especially fundamental level (which, as we saw in Chapter 2, so inspired Whitehead) and the fact that at the level of direct experience natural things maintain their existence as uniform entities; that is, they are held together as unities and as instances of certain types—that is, Forms—by virtue of the good. And if all natural entities are at their core and in point of metaphysical fact configurations of mathematical elements (e.g., triangles, as in the Platonic schematic), then it is not a stretch of language to say that a particular entity is a grouping of mathematical elements constituting an individual instance of a type existing so as to maintain that type-instantiation for the duration of its existence. Furthermore, if "all things" is indeed exhaustively comprehensive, then the good holds together each individual element in mathematics itself, as well as each theorem in mathematics to the extent that a theoretical statement (e.g., in algebra, the binomial theorem or group theory itself) stands as a particular instance of mathematical thinking.

In sum, we are taking the "holding" of all things "together" to produce a multi-dimensional display of order (not without measures of motion as they characterize certain dimensions of reality). By way of a metaphysical contrast of singularly dramatic sort, we might cite "booming, buzzing confusion," William James' famous thumbnail sketch of an infant's untethered experience seemingly inflected with a hefty ingesting of Heraclitus run amok, as it were. In view of this possibility, we shall be especially interested in the particular aspects of order that obtain when mathematics intersects with metaphysics.

All Forms Depend on the Good

The shift from the studied hesitancy in the *Phaedo* regarding the nature of the good and the forthright dependence of the Forms on the good in the *Republic* is remarkable. At *Republic* 509b, Socrates is laying down his position on fundamental reality and he announces a claim that appears to dwell at, or certainly near, the summit of metaphysical speculation—the Forms, presumably all the Forms, depend on the good for their reality: "the objects of knowledge [the Forms] not only receive from the presence of the good their being known, but their very existence and essence [οὐσίαν] is derived to them from it, though the good itself is not essence but still transcends essence in dignity and surpassing power." Glaucon, Socrates' conversational partner at this point, reacts: "Heaven save us, hyperbole can no further go."

How to determine the relation between the good and the Forms? If we take Socrates at his word, this relation involves a kind of unique dependency, although the details of this relation are difficult to state. If all Forms are immutable and eternal (as, e.g., the *Phaedo* maintains), how can they depend on something other than their own modes of existence not only to exist as Forms but also to establish the specific content exhibited by each Form (beauty, justice, etc.)? We will not attempt to articulate the requisite metaphysical depths for answering such a question. We do hazard a point that in context seems safe: Socrates certainly wants to establish the good, however it may be accessed, as at least as real as the Forms even if in the final analysis we sympathize with Glaucon's feeling of being overwhelmed by the special reality of the good as far as the limits of rational discourse are concerned when we attempt to deal with such exalted metaphysical ultimates.

Mathematics: The Good and the Divided Line

In *Republic* VI, Socrates distinguishes between the good in itself and "the idea of the good" [τὴν τοῦ ἀγαθοῦ ἰδέαν 508e2-3]. The latter reality, the idea of good, is that which "gives their truth to the objects of knowledge and the power of knowing to the knower"; this idea must therefore be conceived as "the cause of knowledge, and of truth in so far as known" (508e-9a). Since Socrates says that he is incapable of speaking of the good in itself (506d-e), it is crucial to recognize the relation between the idea of the good and the good as such: This idea stands as "the offspring of the good which the good begot to stand in a proportion with itself" (508b12-13) and, as Socrates proceeds to demonstrate, he does have glimmers of the effects of this reality on knowledge and the objects of knowledge.

We will not speculate on the meaning of "begot" [ἐγέννησεν] in this claim, but the import of the passage is relevant to the essay's argument since the idea of the good, which Socrates is in the process of describing, is asserted to have originated directly from the good and to do so in a manner proportionate to the good itself. Thus, it seems safe to infer that any specific reality or relations between or among realities divined from what Socrates says when considering the idea of the good are also, at some point (although perhaps in a mode of reality necessarily inaccessible to complete philosophical articulation), ultimately due to the good as such.

The offspring of the good is a subject on which Socrates is willing to speak. And since Glaucon wants to know whatever can be known about the good, he eagerly insists that Socrates reveal everything Socrates has gleaned from the idea of the good, the similitude of the good itself. Socrates' initial reply is suitably guarded: "I am omitting a great deal" (509c7). But Glaucon (and the reader) are nonetheless privy to the articulation of a more than significant train of thought.

The Platonic Socrates addresses Glaucon's interest in the good by requesting that he construe the regions of reality according to the dimensions of a line, variously subdivided—this is the divided line, one of the most influential (and discussed) images in philosophical literature: "Represent them then, as it were, by a line divided into two unequal sections and cut each section again in the same ratio—the section, that is, of the visible and that of the intelligible order" (509d6-8). Since the good as such is ineffable, any substantial content attributed to the "idea of the good" must be, necessarily, an interpretation of whatever evidentiary foundation an examination of the good can provide; as a result, we may assume that the idea of the good—existing as our sole access to the good in itself—is open to at least minimal specification.

According to the divided line, mathematical thinking is located one sector below the "highest" available mode of conscious intellection. Furthermore, the objects of mathematical thinking, the metaphysical counterparts to this mode of cognition, are deployed immediately below the Forms characterized as such, in their full canonic sense. It is relevant to note that when Socrates is pressed to say what he can about what, from his perspective, is available concerning the similitude of the good, he introduces a fundamental mathematical element—a straight line. He then operates on this foundational image with additional mathematical elements: cutting the line (thereby generating lines available to be construed as smaller sections of one and the same original line) and doing so according to precise relations generated by these operations, that is, ratios characterized in terms of equality and inequality. Thus, when the Platonic Socrates must evoke whatever is accessible of the good in words, his first instinct is to represent the good according to mathematical specifications, suggesting thereby the intimacy between the good itself (i.e., the idea of the good) and what is available to the philosopher as far as approaching, and approximating, the full (yet in important respects nonetheless still hidden) reality of the good.

Socrates has told Glaucon that much about the good will be left unsaid. We may read this to suggest that at least part of what is unsaid stands as an open invitation to reflect on what has been asserted about reality in the name of the good and to elicit implications from interpretations of those assertions. Whether or not what we claim in this book will be anything Socrates would endorse as a legitimate expostulation of or sequel to the good, insofar as its content can be articulated, cannot be determined. We do maintain, however, that the positions developed in this chapter (as well as Chapter 8) are justified, if speculative, readings of the Platonic vision of the good when viewed from the perspective of how the divided line incorporates mathematical elements and the location of the domain of mathematics along the divided line itself, taken as a schematic unity.

We will develop what stands as, in several respects, a starkly drawn break between the good as depicted in the *Republic* (and anticipated in the *Phaedo*) and the good as it might be envisioned when we give mathematics center stage. As a result, what follows is limited to speculative implications drawn from these textual givens. Indeed, the book is, again (as stated in the Preface and mentioned again later), not an attempt to pursue any of the standard modes of Platonic exegesis.

Commonalities in Mathematics and Metaphysics

What can this presumably axiomatic connection between the good (even in its guise as, in some sense, the derivative idea of the good) and the Forms show concerning the connection between mathematics as such and applying mathematical considerations as a means of informed critical appreciation of metaphysical positions? The answer depends initially on identifying and developing properties, or commonalities, shared by mathematics and metaphysical inquiry.

In Chapter 4, we demonstrated the affinity between the complex structure of Abelian groups and the realities envisioned in the Platonic process of collection and division. The Aristotelian critique of Platonic theorization concerning mathematical entities (Chapter 5) was driven in part by elements sharing this affinity and also by working in conjunction with aspects of Aristotelian metaphysics. Selected features of the *same notions and lines of argument* employed in Chapter 5 were then directed at the Platonic doctrine of collection and division (Chapter 6).

Here are primary instances of these notions: comprehensiveness, oneness, diversification, integrity, totality, class, particularity, unity, property, differentiation, relationality. It may be observed that this list has been compiled from the slate of asterisked terms appearing in Chapter 6. All these terms were instrumental in the Aristotelian-inspired comments, developments, and criticisms appearing throughout the arguments analyzed in Chapter 6. The use of the asterisk ploy was strategic; the intention was to indicate in an indirect manner that the diversity of analyses pursued in this chapter was based on elements with a common locus of origin. For in fact, this set of shared structural factors has been generated by an interplay between analytical and evaluative treatments of mathematical as well as metaphysical positions.

Although the elements comprising this set were derived from a limited context of inquiry—criticism of mathematical philosophizing via the Aristotelian bank of arguments in *Metaphysics* Books M and N and parallel criticisms of a Platonic exercise in metaphysics (collection and division)—we propose to take the interplay as such as a procedural unity with a uniformity of structure. We will call this unity "equiprimordiality." The basic constitution of this unity as a principle may be stated thus: *mathematics and metaphysics share certain general, abstract features; when identified, these features can then be employed as a framework for assessing metaphysical positions.* The arguably cumbrous word "equiprimordiality" is introduced by design, a terminological alert to consider taking seriously a factor that typically receives little emphasis when analyzing and evaluating the content and conclusions of metaphysical positions. In that regard, we should also observe that the shared properties listed above are found in treatments of a wide variety of metaphysical investigations, a dimension of common reality that will be explored presently. But it is crucial to recognize for now that working mathematicians function at a level of metaphysical determination even if this determination remains unvoiced, awaiting articulation in a form apart from the systematic, theoretically driven and defined symbolic dimensions of mathematical expressions taken as such.

Earlier in the chapter, we saw Whitehead's essay on mathematics and the good emphasize the fact that mathematics was *a*, if not *the*, primary avenue toward the recognition of patterns as fundamental features of reality. The notion of commonalities is certainly influenced by this approach to patterns, but we have here extended the scope of patterns by considering abstract features common to *both* mathematics *and* metaphysics. The question then arises: can the functionality of equiprimordiality, the interactive relevance of mathematics and metaphysics as filtered through a list of suggested properties, be generalized so that mathematical theorizing becomes structurally aligned with other metaphysical positions (not only in Plato but by any metaphysician)? In our inquiry at this juncture, the notion of equiprimordiality derives, as noted above, solely from reflection on the fluidity of application to (a) Abelian groups and (b) Plato's position on collection and division. In the remainder of Chapter 7 as well as through Chapter 8, we argue for a much more expansive range of relevance.

A guiding motif of this book has been the notional affinity between the grouping aspect central to Abelian groups and the process of grouping as it constitutes the collection phase of Platonic collection and division. Now if the group structure of collection and division as a methodology to secure metaphysical conclusions is analogous to the group dimensions of other metaphysical positions, then the speculative inference may be drawn that if the properties employed in the critical evaluation of collection and division are sufficiently general, they may be similarly applicable to *any* metaphysical structure. The metaphysical dimensions underlying the structure of mathematical positions, considered by themselves, can therefore be introduced as heuristic media for the purposes of critical analysis of metaphysical structures presented as such. The heuristic factor here is especially important. What we are proposing is that the metaphysician would be examining what can be learned about metaphysical positions, all of which are analogically groups of a sort, when these positions are confronted with a set (or selected members of a set) of abstract notional considerations derived from mathematical sources.

This kind of application involving juxtaposed commonalities and the derivation of consequences such exercises may—or may not—engender does not entail eliminating foundational differences between mathematics and metaphysics. The point is rather that substantive differences between the two disciplines do not affect the commonalities, all involving generalities of high order, characterizing both metaphysics and mathematics. Careful examination of such applications may therefore be valuable in eliciting structural, systemic considerations pertaining to the proposed metaphysical structures. It is perhaps evident that vestiges of Platonic realism are embedded in this approach; given appropriate conditions for eliciting their discovery, we may assume that the systemic features under scrutiny were in some sense already "there" to be discovered.

We propose to extend applications of equiprimordiality beyond the context of Abelian groups in conjunction with collection and division. In order to establish this possibility, one way to proceed is by returning to Plato, in particular the account of the good in the *Republic*. The good, if approached from a certain perspective and sufficiently articulated, reveals significant commonalities between mathematics and metaphysics. In fact, the range of these commonalities would extend the limits of these notions significantly beyond those derived from the intersection of Abelian groups and collection and division. *All* these commonalities would then become available for use in the critical assessment of purely metaphysical positions.

It should be emphasized that the argument at this juncture of the essay does not require endorsing all dimensions of the good identified in the *Republic*. The only requirement is an impetus to take seriously for metaphysical purposes some (or all) of the notions listed above as "commonalities" and to seek clarity in the predication and application of these realities as far as producing the most satisfactory metaphysical accounts. The "gathering" of these notions as an integrally coherent and blended set may or may not be found accommodating by a working metaphysician, especially at first glance; however, the claim here will be that this confluence of generalities provides a broadly relevant heuristic schematic for the critical assessment and possible expansion of any given metaphysical position.

Mathematics and the Divided Line—A Phased Reformulation

The interpretive strategy in this section is straightforward: we propose to preserve the divided line intact as described in the *Republic* and with the same subdivisions, but to redefine its function, concentrating on *the segment of the line devoted solely to mathematical matters* (i.e., realities as well as properties of these realities). This reconfiguration will show how the good can identify and coordinate properties which, given their fundamental character, are sufficiently malleable to apply in mathematical as well as metaphysical situations.

The *Republic* account of the divided line intimately juxtaposes the domains of mathematics and what we would now identify as metaphysics, especially pitched at a high level of abstraction. If, generally speaking, the Platonic vision of the fullness of reality is at all theoretically credible, neither mathematics nor metaphysics is skewed in any fundamental way by the proposed reformulations and mutual exchanges of inquiry and scope. The ascent up the line's hierarchical divisions situates mathematics so that it contributes to awareness of what can be construed at the apex of metaphysical reality; and if we allow traversal of the line in the opposite direction, then we may expect at least a measure of the reality proper to the Forms to infuse the level of reality situated directly beneath the Forms, within the world of mathematics. The approach to the good argued for in this book preserves, but in a number of ways significantly enhances, these broad procedural and metaphysical strokes. Indeed, the metaphysical content of mathematical elements, even if its degree of reality is in some sense diminished when compared to that of the Forms, retains sufficient metaphysical depth, so to speak, to function heuristically with respect to any explicitly stated metaphysical position.

The reconfiguration of the good will be executed in three phases; the culmination of this analysis will be proposing a speculative unity that consolidates the conclusions of the three phases. This diversified unity will be named the "transcendental grid." As we shall see, this device encompasses the possibility of establishing an unlimited number of interpretive schemes introduced as a speculative reformulation of the divided line insofar as this configuration has been informed by the good.

Realities and Properties [First Phase]

We will preserve a standard representation of the divided line (drawn vertically); we will also retain the sharp distinction (afforded by the fact that a line can readily be subdivided) between different sectors of reality, that is, Forms on the "top," mathematical realities "below" the Forms, physical objects and their reflections at the "bottom." Forms and mathematical entities occupy different segments of the line. According to the *Phaedo* and the *Republic*, the good holds together all things, and all Forms depend on the good for their very existence. Justice and beauty are both Forms; they also differ from one another. However, as Forms, justice and beauty share properties that characterize all Forms. Also, if Forms as such depend on the good, then we may infer that all properties inherent in Forms depend on the good, both the single property constituting a given Form's uniqueness and all the properties common to all the Forms.

As the initial phase in the reformulated good, we suggest distinguishing between (a) the substantive content of Forms (which is singular and also unique to each Form) and (b) the structural characterizations of Forms (which are plural—immateriality, immutability, etc.—and thus common to all Forms). Concisely stated: we propose separating *formal* properties from *substantive* properties; thus, the reformulated segment of the line is constituted by properties of realities, not realities as such.

This distinction has two immediately useful consequences. First, Socrates maintains in the *Republic* that any mathematical entity cannot be grounded according to a hypothesis in the same way as any reality derived from participation in a Form. Scholars disagree on the precise rationale for this important difference. The dispute revolves around (a) the status of mathematical entities with respect to their mode of reality and (b) formulations pertaining to the modes of generalization on the basis of which mathematical claims are derived. This interpretive dispute is, however, effectively obviated if we assume that (a) *all properties* pertaining to Forms as the "highest" kind of reality can be considered apart from the Forms themselves and that (b) these properties will also characterize the proper subject matter of mathematics. This conclusion will follow regardless of whether the reality of mathematics as represented by its stretch of the divided line is construed as a collection of distinctive entities (the type of reality directly "below" the Forms) or as hypothesized generalizations of those entities. Again, the suggested reconfiguration of this segment of the divided line concentrates on properties of realities, not the realities themselves.

The first phase of the reformulated divided line involves replacing distinct mathematical realities—geometrical figures (circles, triangles) and numbers—

with properties. This substitution may be permitted if separating Forms from the properties of Forms is accepted as a matter of metaphysical principle. This separation also sanctions the second useful consequence. For if the properties of realities can be isolated and then studied as separable features of the highest kind of reality, then we may suppose that, in general, the examination of the properties of realities can become a worthwhile endeavor in its own right. Thus, if *all* realities "below" the Forms (as well as mathematical realities) derive their modes of reality from the Forms, then the properties of these realities are also eligible to be considered as isolatable aspects, even if these properties inhere in realities other than Forms. We will identify instances of these properties in due course.

Mathematics and History [Second Phase]

The history of mathematics is a fact. It is known that non-Greeks were doing mathematics in the third millennium BCE. Plato knew geometry (recall its use in the *Meno* and *Timaeus*) but it is generally acknowledged that the initial formulations of, for example, spherical geometry did not appear until after Plato (and non-Euclidean geometry was not a mathematical reality until far into the future). It appears unlikely that Plato envisioned mathematics as having a history in areas of investigation other than what the dialogues assert about number theory and plane geometry; these types of mathematics had existed for an extended period but not in the sense that their appearance would have been located on a timeline of sorts.

The second phase of the reformulated segment of the divided line renders it inherently temporal. This fundamental dimension of temporality will accommodate the evolutionary nature of mathematics, that is, the virtually constant development of subject matter and material *within* the domain of mathematical realities and along its historical dimensions. Furthermore, the effects of the good according to this reformulation are, in an important sense, transhistorical—thus, the ramifications of the good generated by triangles in Plato's thought were historically *futural* in the fact of their import extending to areas of investigation pursued (by Whitehead) only millennia later. But the effects of the good can also be construed as historically *retroactive* in the sense that what has been asserted and derived from the good in contemporary modes of metaphysical inquiry can also find antecedents in past metaphysical reflection and resultant theorized positions.

In the reconfigured segment of the divided line, we are envisioning the good to embrace a necessarily historical dimension in order to accommodate the crucial importance of the history of mathematics and the ongoing and palpably recognizable continuity of fresh discoveries and advances in virtually all areas of mathematical investigation. The affinities between Abelian groups and Platonic collection and divisions are not isolated phenomena; indeed, mathematics functions throughout its history with respect to a foundational consanguinity with metaphysics in a manner similar to the history of metaphysics taken solely by itself, as an ongoing series of autonomous and self-limited theoretical ventures. To separate time and history from the divided line's mathematical segment is to compromise, if not seriously impede, the myriad vibrant possibilities whereby mathematics creates out of its own creations.

Mathematics and Continuity [Third Phase]

The Platonic use of a line to represent the range of reality, from the Forms to shadows and images of particulars, has been praised for its schematic simplicity and criticized for its apparently irresolvable implications. In the third and final phase of the reconfiguration of the good, it may therefore be relevant to note that nothing new is added to the structural character of the line; rather, we consider here implications derived from assuming that a line is, as such, *continuous*. The continuity of the line can be understood to represent the common thread of the good permeating all major subdivisions of reality, including everything that falls under the domain of mathematics. The relevant aspect of continuity is borrowed from *Physics* Book III, where Aristotle shows that if a line is continuous, it can be divided and subdivided apparently without limit. As a result, the reconfigured segment of the divided line will exhibit the possibility of this limitless subdivision from end to end, from its upper limit (facing the Forms) to its lower limit (abutting the concrete reality of material things).

So far, we have interpreted the line so that its content (a) is characterized by properties and (b) reflects the fact that mathematics has its own history. When a given type of mathematics appears in history, establishing itself as a standard form of inquiry, this type intersects with the line. The intersection emerges along the line at a point determined by when that type appeared in history. We will assume that the historical origin of mathematics as a mode of human behavior is located toward the "low" end of the line, with the passing of time constituting history represented by movement "upward" toward the high end of the line. If so, then arithmetic and geometry intersect the grid at low levels; by contrast, Abelian group theory connects with the line at a higher point, presumably above geometry. Calculus, for example, will occupy a position somewhere above geometry and somewhere below Abelian groups.

For purposes of the argument in the essay, exactly where each area in mathematics intersects the line is secondary to the fact that at the point of intersection, each and every type of mathematics displays a set of realities, that is, properties of a high order of generality as they characterize that type of mathematical inquiry. Mathematics has evolved and continues to evolve. The properties produced by these intersections with the line should be examined to determine how they might be juxtaposed with avowedly metaphysical positions, thereby facilitating more advanced critical attention to those positions. As commonalities derived from mathematical elements and proof structures, these properties are capable of disclosing critical and expansive insights into mathematical positions. The challenge for the metaphysician becomes to appreciate the *heuristic value* generated by these insights, absorbing and analyzing what the mathematicians are "saying" about the structures of their patterns that can be "translated" into vehicles of metaphysical expression using appropriate vocabulary, whether delivered in ordinary language or through freshly coined technical terms.

Two additional facets proper to this aspect of the reconfigured line must also be described. First, since the line is construed as continuous, an unlimited number of novel mathematical positions are possible (reflecting the infinite divisibility of a line). One, some, or all of these areas of mathematical investigation may, by virtue of the principle of equiprimodiality, contribute properties capable of illuminating a given instance of metaphysical inquiry. In this regard then, there is no single limit to "pure" mathematics and, by extension, no limit to the ways in which advances in mathematics can contribute to the development of the metaphysical enterprise. Second, the singularity of the line (in the standard illustrations of the divided line) is a diagram solely for the sake of visual convenience. The history of mathematics reveals in dramatic and graphic fashion that as far as content is concerned, mathematics has and doubtless will continue to splinter its domains into areas, sub-areas, sections of sub-areas, and so on. In effect, there are an unlimited number of divided lines (i.e., the mathematical segments thereof), each representing an advance generated from a previously established mathematical position and thereby becoming another locus of many additional advances. It is not an interpretive stretch to anticipate the real possibility of radical innovations to foundational components of mathematics as well as expansions and extensions of areas of current investigation.

Constant throughout all these changes and modifications are properties displayed by both sectors of reality, that is, commonalities; by contrast, variability characterizes the extent to which mathematical entities are and become directly relevant to counterpart metaphysical groupings (whether systematic or regional). An important corollary: As discoveries in mathematics become increasingly abstract, it seems reasonable to assume that their potential applications to metaphysical configurations may become correspondingly difficult to discern. However, we cannot (or, perhaps, should not) automatically assume that future structures of mathematical entities, formulations, and theorems will become, by themselves, so impenetrable as to preclude the very possibility of relevance to future metaphysical speculation. Mathematical imagination's place in the process of such philosophizing (as Whitehead will point out in Chapter 8) exercises a powerful capacity for insight into and control of seemingly complex mathematical phenomena.

The Transcendental Grid—An Intersection of Mathematics and Metaphysics

This concluding section introduces the "transcendental grid," a narrative device intended as a technical term to replace the Platonic reality, the good, as that reality appears in the Platonic dialogues discussed in this chapter. The transcendental grid represents a reconfiguration of the divided line, including a complex array of factors embodying the underlying assumptions discussed in previous sections of Chapter 7. The transcendental grid encompasses elements of mathematical and metaphysical dimensions which can be traced to the comprehensive scope of the good as an ultimate foundational reality—assuming that the good in its Platonic expression can be rethought and restated according to the suggestions made above.

Indeed, the specific historical matter—whether this position can be legitimately derived from the Platonic texts on the good—has in a sense become immaterial. The good as reconfigured in this book has in effect become a "transcendental medium" within which correlations between mathematical and metaphysical elements, construed at the level of theory, derive their origin and capacity to interact with one another. The relevant point then is not where and how the approach to mathematics as constitutive of metaphysical considerations and culminating in the principle of equiprimordiality originated, as far as available texts in philosophy are concerned; rather, the explanation and description of this intimate connection between mathematics and metaphysics should be judged on its own merits. If, in the final analysis, equiprimordiality is deemed to be without purpose with respect to the metaphysical enterprise, then how it originated is a matter of small consequence. However, the argument in Chapter 8 will reinforce the fact that equiprimordiality is of considerable value in enhancing the results of metaphysical inquiry.

To summarize: at the outset of this chapter, we asked whether it is possible to expand the range of the kind of critical analyses illustrated by the interchanges between Abelian groups (as an area of mathematics) and collection and division (as a position in metaphysics). The answer is in the affirmative, and the transcendental grid is a foundational basis for such critical and evaluative exercises.

Here is the transcendental grid in schematic outline:

(i) the transcendental grid in its entirety is derived from that section of the *Republic*'s divided line intended to represent mathematical reality (as distinguished from the "higher" section of the line encompassing only Forms and, toward the other end of the line, the "lower" section representing material objects);

(ii) however, this section is not characterized by typical mathematical elements—numbers, planes, circles, and so on—taken as self-subsistent unities but solely by *properties*, pitched at a high degree of generality, insofar as they can apply to mathematical elements. This aspect of the transcendental grid obviously stands as a crucial step in the reformulation of the doctrine of the good in the *Republic*;

(iii) furthermore, these properties inform the *entire continuous segment* of the line representing mathematical dimensions of reality;

(iv) the line is marked by intersections representing the emergence of distinct types of mathematics, the locus of each such intersection occurring within the limits of the line (i.e., at a point somewhere between the line's terminal demarcations). As a result, different mathematical areas will, as unified contexts of investigation, intersect the line at different, definite locations;

(v) finally, each of these intersections includes a specific set of properties determining the metaphysical character of the type of mathematics evoked by the given intersection, thereby instantiating the principle of equiprimordiality. (The constituent set of properties for Abelian groups has been identified above.)

As noted, we have named this set of factors (including the underlying assumptions discussed in this chapter)—a unified vehicle of analysis incorporated within the reformulated approach to the good—the transcendental grid. The grid stands as a reconfigured and speculative approach to the structural features (and implied metaphysics) of the divided line in the *Republic*, as well as a source for demonstrating the illuminative power of mathematical areas of inquiry when juxtaposed with metaphysical ventures.

Here are the essential differences between this grid and parallel positions, that is, segments along the canonic divided line. The transcendental grid is (a) historical, (b) continuous, (c) entirely conceptual, and (d) initially heuristic when intersections on the grid are applied to metaphysical accounts. The name, "transcendental grid," thus evokes the basic purpose of the essay with respect to the interplay of mathematics and metaphysics. "Transcendental" is a direct nod to the Kantian use of the term but reconfigured in the present setting to refer to "conditions for the possibility of revising a given metaphysical position or system." Thus, "transcendental" points to stipulative conditions but not as they function for Kant—that is, a necessary prelude to the very possibility of experience itself; rather, more narrowly, the term indicates the possibility of deriving properties that identify conditions for the possibility of experiencing theoretical connections between mathematics and metaphysics. These connections can be consolidated and articulated as notions of high generality, thereby becoming procedurally useful for applying the illuminative power of mathematical areas of inquiry toward questions and concerns at key junctures of metaphysical inquiry.

Due consideration from this perspective thus makes it evident that Abelian group theory constitutes one of the intersections (introduced as "iv" in the above outlined description) as a factor in the constitution of the transcendental grid. Indeed, the concerted attention to elementary group theory as embodied in Abelian groups reaches its most advanced point, in terms of the speculative goals of this book, when we consider this area in mathematics as an instantiation, a realization, of the transcendental grid. This area of group theory and the associated discussions of its influence in the intersection of mathematics and metaphysics represent a paradigmatic realization of this grid. Furthermore, and with the benefit of narrative hindsight, we are now in a position to recognize that treatments of the metaphysically driven critiques of Platonic philosophy of mathematics (Chapter 5) and subsequent assessment of the Platonic position on collection and division—using the same metaphysical elements as critical instruments (Chapter 6)—illustrate, as a conjoint venture, how a given realization of the transcendental grid works. In fact, we can show that the transcendental grid grounds the possibility of an unlimited number of additional such intersections, some or all of which may lead to incisive applications in the domain of metaphysics.

Hindsight is also useful in another, related theoretical context. The array of elements identified earlier in this chapter (in "Commonalities in Mathematics and Metaphysics") illustrates one set of determinate factors emerging from the intersection represented by Abelian group theory as a realization—

one realization—of the transcendental grid. This gathering of properties is an example of (v) in the schematized statement of the transcendental grid outlined above. We mentioned the juxtaposition of Chapters 5 and 6 in order to identify the origin of a set of seminal elements found in critical examinations of both the philosophy of mathematics and metaphysics proper; now, however, we can appreciate the fact that the appearance of this interplay depends on more fundamental and more inclusive approaches to the realities of mathematics and metaphysics, approaches produced by situating these exercises in transcribing reality according to broad-based parameters derived from speculative developments of the Platonic notion of the good.

Additional explanation, argumentation, and defense are required in order to solidify the potential relevance of the position reached at this point. We may nonetheless be allowed to emphasize as a watchword of philosophical procedure that the more influence the metaphysician allows mathematics to exert on a metaphysician's worldview, the more this metaphysician will at least want to investigate the possibilities derivable from the application of the transcendental grid as far as developing the best accounts possible in a given area of metaphysical analysis. To justify that claim is the primary burden of Chapter 8.

Notes

1 This book approaches "the good" from a substantive metaphysical perspective. For a clearly stated analysis of "the good" from the perspective of modern ordinary language philosophy (which in turn leads to markedly different philosophical conclusions in the context of how "good" functions in value theory), see the chapter "The Meaning of 'Good'" in Sir David Ross, *The Right and the Good* (Oxford: At the Clarendon Press, 1930), 65–74.

2 Alfred North Whitehead, "Mathematics and the Good," in *The Philosophy of A. N. Whitehead*, 2nd ed. (LaSalle, IL: Open Court,1999): 666–81. Quotations from the essay (here, 666) will be cited in the body of the text. We may observe that at one point, Whitehead neatly evokes, in summary fashion, the underlying rationale for much of the material from Plato and Aristotle analyzed so far in this book:

"The doctrine of number from the Greek period onwards has always included queer little contradictions which thoughtful people disregarded" (671). We have attempted herein to regard thoughtfully some of the logical and theoretically problematic aspects of the philosophy of mathematics at its historical point of origin, especially from the standpoint of what can be learned thereby that could be useful for metaphysics.

3 The words "Whitehead" and "whimsical" would not typically be seen in the same sentence, especially in a discussion of an essay on mathematics and the good; however, we read the following (perhaps apocryphal) anecdote regarding the history of algebra (677):

> Of course, algebra grew slowly. For centuries it was conceived as a mode of asking for the solution of equations. Somewhere in medieval times, an unfortunate emperor, or other bigwig, together with his court, had to listen to a learned Italian expounding the solution of a cubic equation. Poor men—a lovely Italian afternoon was wasted!

4 The good appears again in the *Statesman* and is discussed at some length. However, determining the function of the good in this dialogue depends on how the structure of the dialogue is interpreted as a narrative and discursive unity. If therefore the dialogue is taken as primarily an aporetic exercise, then any substantive claims about the good must be examined with care. For such a reading, see: D. White, *Myth, Metaphysics and Dialectic in Plato's* Statesman, 131–94. Similar interpretive caution should be exercised regarding the complex discussion of the good at the conclusion of the *Philebus*.

5 A. E. Taylor cites this passage but apparently without recognizing its full implications regarding the fundamental connection between the good and the Forms. See A. E. Taylor, *Plato: The Man and His Work* (New York: Meridian Books, 1956), 287.

8

Mathematics and the Good

On the Maximization of Metaphysics

Abelian groups of mathematics and the metaphysical agenda characterizing Plato's later interest in collection and division: initially, these structures might appear to be unrelated. However, the argument of this book, starting from the initial presentation of the reality of numbers in the *Phaedo*, has shown that extensive affinities link these exercises in theorizing. Whitehead's neatly divergent treatments of Platonic philosophy of mathematics, coupled with Aristotle's extensive, often finely grained critiques of his teacher's positions on the reality of numbers (and the subsequent application of selected aspects of these critiques to the methodology and substance of later Platonic metaphysics) contribute force and direction to a series of shared commonalities—abstract, general notions that characterize mathematics as well as metaphysics. These commonalities function in ways captured by what we have called (in Chapter 7) the principle of equiprimordiality. This principle, in turn, is instrumental in leading to the establishment of the transcendental grid and the affirmation of Abelian groups as an important illustration of the grid's function. Additional discussion of the structure of the grid, as well as further applications of its use and some final reflections on beauty, imagination, and adventure as resident features in applications of the grid, constitute the main themes in Chapter 8.

Every dimension of the book's general argument has been rooted in Plato, Aristotle, and Whitehead. However, the results of this investigation are entirely those of the author. We have considered and developed texts from these thinkers employing a number of penetrating sources on mathematics as critically analyzed from a metaphysical standpoint. Chapter 7 concluded with a description of the transcendental grid, a speculative unity defined by an array of elements broadly cast and ambitious in potential application. The identification of a slate of realities animating this unity—working examples of the principle of equiprimordiality—stands at the culmination of the book's substantive goals. The results of our inquiry underlie positing a series of abstract notions and conceptual schemes derived from the intersection of mathematics and metaphysics. These realities are capable of being employed for purposes of either (indeed, perhaps both) critically evaluating or expanding the substantive range of given metaphysical positions.

The overall argument leading to the statement of these notional and schematic elements is speculative, but not without tendering suggestions in terms of possible applications to "real world" metaphysical positions. However, the full range of speculative activity has yet to be stated, and this concluding chapter of the book will develop various aspects of this decidedly practical theme. We will outline ways to apply a particular instance of the transcendental grid and thus recognize the metaphysical yield when mathematics intersects with elements derived from a foundational and determinate approach to the good. We will also indicate, in a general sense, the effective boundaries of metaphysical inquiry when limits emerge as a result of pursuing such applications. The pivotal importance of imagination in the processes of application will be underlined, that is, imagination understood as the personal prism for appreciating the presence of, and tendering substantive results from, the fundamental interaction between metaphysical thinking and mathematical theorizing. In this setting, Whitehead returns to provide suggestions for employing imagination as a primary medium of process for integrating the commonalities defining the two disciplines. Whitehead also offers a concluding perspective on the adventurousness permeating all such thought—a mode of articulating the structures of reality on a level of high generality yet an endeavor not without certain measures of philosophical risk.

Is There a "Best" Metaphysics?

Plato, Aristotle, Whitehead—all are preeminent in the general area of philosophy known as metaphysics. However, their respective positions on the nature of reality differ considerably from one another. And obviously many more theoretical positions could be added from philosophers who have reflected seriously on questions pertaining to what is, and is not, real. As a result, the question posed as the heading for this section may appear, if not logically suspicious on the face of things, quixotic at best. Some, perhaps many, will reject the question about a "best" metaphysics out of hand as a facile manipulation of inherently amorphous terms. But a measure of sympathy for this book's project will invite at least attempts to clarify the scope and purpose of the question. After all, our basic argument has culminated in proposing a table of concepts, rubrics of a sort, which have brought the good into more manageable regions of analysis.

Initially, perhaps, the "best" metaphysics would appear to be whichever system (or articulated position on a given issue) seems to square with one's intuitive—and considered—reactions to this account (or these accounts for students interested in multiple and distinct metaphysical matters). Crudely put then, reality from a metaphysical perspective is what justifiably and logically exists in the mind of the person thinking about reality. But we have argued that it is possible to move beyond this self-determining dimension of metaphysics by the analysis and development of a reconfigured approach to the good as embodied in the structure and elements of the transcendental grid. As a result, we suggest that a first step in establishing a connection between the transcendental grid and the good regarding the possibility of the "best" metaphysics rests on a simple realization. From the standpoint of basic grammar (and in a context of linguistic immediacy far removed from high flights of speculation), "good" in its adjectival sense has degrees, that is, "better" and "best." This handy and elementary subdivision suggests a line of analysis consonant with our speculative adventure.

The answer to the question at hand involves a series of interpretive reflections derived from the Platonic vision of the good. Thus, the best metaphysics will mean, in the present context, that metaphysics which incorporates as fully as possible determinable aspects of the good, aspects which are as specific as the investigator of the good can make them. For example, if we pay attention to the Platonic position, one approach to the best metaphysics is that it will be the most comprehensive (reflecting how the good is fundamentally related to the Forms and then to everything, all particulars, caused by or in some way dependent on the Forms). This approach suggests that comprehensiveness in metaphysics must be concerned with the totality of what exists. We shall, however, for present purposes take "comprehensive" in a circumscribed sense, that is, to be relative to the scope of metaphysical inquiry as indicated by the texts under scrutiny. If the scope is *systematic*, as it is from the work of Parmenides through Aquinas and into the relatively recent metaphysical past—Bradley, Dewey, Sartre, Quine, Badiou, and others—then the best comprehensive metaphysics will be one that maximizes all relevant characteristics of that system given its axioms and experiential points of departure; everything that exists will have a place in the theoretical articulation of totality, given available levels of discourse and notional configuration. If, by contrast, the scope of inquiry is *regional* and therefore restricted by the metaphysician to a given issue, region, or concept in metaphysics, then the "best" metaphysics will be one that establishes as many characteristics relevantly pertaining to that particular subject matter as can be determined within available textual, theoretical, and experiential parameters. In this more limited respect, "comprehensive" means everything relevant to a complete account of a given segment or sector of the metaphysical enterprise.

To affirm that an instance of metaphysics is "the best" is therefore to appeal to the factor of comprehensiveness in the senses just specified. But two immediate caveats: first, saying that a metaphysics is "best" is not saying that it is unassailable, impervious to any and all criticism; any metaphysics promising to have achieved a level of comprehensive analysis must also be prepared to defend itself against any reasonable criticism. Nor, second, is it to say that there may not be more than one "best," depending on the development of detail based on a set of given elements. Two systematic metaphysics may, for example, both be best if they are derived from the same set of originating elements, generate the same explanatory range but do so by differing routes of development and internal linkage between and among each system's fundamental components. Thus "best" in the context of our inquiry does not refer to an account that is singular in form and content and without parallel when juxtaposed with different (in some respects) articulations of reality.

These viable possibilities may seem to suggest that an ascription of "best" to a given metaphysical account becomes tantamount to an honorific descriptor, doing nothing more substantial than evoking the theoretical relevance of comprehensiveness as that characteristic is embodied in the Platonic version of the good. From a standpoint of theoretical practicality then, it seems that there is no "best" (i.e., single best account in all respects) metaphysics or, at most and at the edge of logical possibility, that such an account awaits its revelations.

But the significance of the honorific use of "best" does not begin and end in a purely linguistic condition of semantic indeterminacy. For even if there is no best metaphysics in any single, justified sense, reflection on the ordinary grammatical usage of degrees of adjectives—positive, comparative, superlative—may intrigue interested metaphysicians to examine the products of their labor and, after sufficient review, to ask themselves: "Can this instance of metaphysics be made *better* than it currently is?" In other words, "best" in, for example, the context of systematic metaphysics would not denote (assuming such a "best" account could exist at all) a self-subsistent, voluminous statement of seemingly limitless speculation liberally laced with a slew of grandly general terms (e.g., and no disrespect intended whatsoever, variations competing in articulated grandeur with the speculative vastness of *Die Wissenschaft der Logik*). To affirm, however, that something is better than it was differs fundamentally from claiming that this something (whatever the something might be) is, in point of metaphysical fact, the best. Therefore, regardless of the style of thought or terminology in which that thought is expressed as constituting a given instance of metaphysical inquiry, we may surely wonder: is it possible that an existing analysis and description of reality, expressed as it must be in concepts and language of some sort, could be made better than it is in its current public incarnation?

If we agree to answer yes, the question then becomes: how can a given metaphysical position be made better than it was? In terms of the parameters of this book, the answer depends on applying that dimension of the good which, we have argued, is embodied in the transcendental grid. Thus the "best" metaphysics—which we now understand to refer to a metaphysical position that is better than what it had been—will be one that includes (but of course is not limited to) articulating commonalities pertaining to intersections between mathematics and metaphysics as indicated along the transcendental grid.

And at this juncture, we are justified in realizing this end by employing the commonalities animating the intersection of the structure of Abelian groups with the Platonic account of collection and division. These commonalities (identified and discussed briefly in Chapter 7) are also prevalent in much of the metaphysical analysis proper to Western thought. The function of these commonalities will now be described and developed as a programmatic ensemble, intended to set in relief the substantive results of the book. This survey stands as a prelude to determining the utility of these results when applied for purposes of producing better metaphysics.

Mathematics and Speculative Metaphysics: Process and Product

The transcendental grid is a tightly reconfigured and bounded stretch of the original divided line in the *Republic*; as such, the grid is devoted entirely to mathematics. The transcendental grid represents a locus for determining the extent to which metaphysical considerations characterize areas of mathematical inquiry. As argued in Chapter 7, the intersection of Abelian groups with the grid emerges as a paradigmatic instance of how this reconfiguration occurs, and what follows from this intersection in terms of equiprimordiality. If we then review the discussion of Aristotelian-inspired criticisms of Platonic positions regarding collection and division (Chapter 6), we will observe how interpretive points made and questions posed resulted from the use of certain elements as they occurred under the aegis of equiprimordiality. Our conclusion: these critical reactions to the Platonic position have, from a plenary perspective, made that position better than it was (from a methodological and metaphysical perspective) when viewed exclusively from the perspective of passages in the dialogues cited and analyzed in some detail in the book.

We now initiate a general account of the processes involved in producing better metaphysical accounts by offering a list and description of basic elements as well as procedures applicable to realizing this end. The list will initiate a four-phased analysis of the process if viewed as a unified and organizationally coherent program—that is, Transcendental Elements, The Heuristic Dimension, Continuity and Limits of Inquiry, Strategies of Application.

Transcendental Elements

The following metaphysical factors, identified as "transcendental elements," encompass items appearing in the section (Chapter 7) "Commonalities in Mathematics and Metaphysics." The following list is ordered, with headings:

M-Principles	*M-Categories*
Comprehensiveness	Totality
Oneness	Class
Diversification	Particularity
Integrity	Unity
	Property
	Differentiation
	Relationality

The constitution of this list results from sustained inquiry. Chapter 4 displayed the affinity between the structure of Abelian groups and key passages in the Platonic theory of collection and division; Chapter 5 presented a varied selection of Aristotle's criticisms of Platonic philosophy of mathematics, and Chapter 6 showed how these criticisms could be adapted for use in modes of analysis and assessment in the context of collection and division, a method of inquiry devoted to metaphysical goals. Chapter 7 then codified the results of Chapters 5 and 6 as a conjoint venture, constituting as such a realization of the transcendental grid as a foundational ploy based on the metaphysical pervasiveness of the Platonic vision of the good and its incisive inclusion of mathematics as a segment of the divided line (with the line, given its original placement in the *Republic*, understood as a complete imagistic transcription of the ranges of reality).

As such then, the list of transcendental elements enumerates primary modes of differentiating the primordial universality of the good in metaphysical settings. Some properties are derivable directly from the Forms (the "highest" elements of generality available from the original divided line); other properties are identified indirectly by envisioning a context of generalities characterized by the kinds of contrasts emergent from a complete system containing, at a fundamental level, Forms as well as particular instances of Forms as we experience these instances in the world around us. The list originates in analytical accounts dealing with mathematical realities; however, given their generality and pervasiveness in Western thought, the elements on the list are also suitable for application to many divergent strains of and approaches to metaphysical analysis.

M-Principles. Principle is understood in the rudimentary realist sense of "that from which anything flows in any manner whatsoever." The prefix "M" invokes the fact that the principles under scrutiny arise from a pair of distinct disciplines, each beginning (in English) with the letter "m"—metaphysics, mathematics. These principles may be readily recognized as a paramount illustration of equiprimordiality; as such, they are umbrella-type designators, that is, extremely broad and inclusive specifications that contain many extant and possible instantiations and developments. Our contention is, however, that each principle's referential range is such that no one principle can be reduced to any of the other three principles.

The following is an elaboration of each principle, briefly indicating its scope when it is considered as a possible mode of application to a given exercise in metaphysics:

Comprehensiveness—how much reality has been introduced into the account, including modal dimensions of possibility and necessity;

Oneness—the sense (perhaps only intuitively felt but nonetheless palpable) of how the factors in the account hang together or fit with one another as a theoretical and explanatory uniformity;

Diversification—the range of differences characterizing the oneness of the account;

Integrity—logical cohesiveness (defined in a standard formal sense) as it pertains to each and every factor in relation to every other factor in the account viewed as a uniformity.

M-Categories. The distinction between M-Principles and M-Categories is more one of degree than kind. M-Categories move from levels of very high generality "downward" toward the concrete reality of individual things. The allusion to Kantian procedure is, again, not to be entirely dismissed as relevant to the purposes of this slate of elements. Each category contributes an aspect of organization and specificity to what metaphysicians typically refer to as the structure (or structures) of reality. Again, as with the M-Principles, we include a brief description of each M-Category:

Totality: everything that exists (everything taken individually, one by one);

Class: type designations (in the broadest sense of the classificatory function of a type);

Particularity: this or that individual entity (taken as autonomous);

Unity: the sense (or senses) in which each particular is *one* thing;

Property: determinable ways describing how one thing (or a single group of things) exists;

Differentiation: ways of distinguishing one thing from any (and every) other thing;

Relationality: how two (or more) things are connected (in the broadest sense) to one another.

These descriptions of the transcendental elements, admittedly spare, are also sufficiently delimited so that the elements, when applied, can be inclusive and flexible. As a result, when any of the elements are juxtaposed with given instances of metaphysical inquiry, they will effectively accommodate as well as direct an interest in making the target metaphysical positions better than they have been.

The Heuristic Dimension

The transcendental elements are derived from the transcendental grid, in particular the intersection of Abelian groups with the resident properties constituting the (now reformulated) mathematical section of the divided line. Their purpose, when duly applied to a given metaphysical exercise, is to contribute to improving the reasoned results of this exercise, to make it "better" in relevant ways than it had been. The transcendental elements listed above will function in a manner sufficiently broad and incisive to be employed as components in the critical evaluation of a variety of extant metaphysical positions. It is vital, however, to understand what might be called the locus of improvement, that is, the general context underlying the possible results of applying the transcendental elements.

We have already made use of "heuristic" at several prior junctures to indicate a certain interpretive approach to various items in terms of their function, given the general thesis of the book. The word "heuristic" comes from the Greek verb εὑρίσκειν, which basically means "to discover." When two metaphysical positions are juxtaposed, either according to equivalent function or with identical or closely similar terminology, the investigator can discover by examining the element juxtaposed with respect to proximate elements in either position how the original element changes, or can be changed, in a way or ways that increase the explanatory power of the originating element.

The "M" affixed to headings of the transcendental elements thus also indicates "meta-", meta-principles and meta-concepts, that is, fundamental elements intended to function about or with respect to principles and concepts as these aspects characterize a given metaphysical account. Meta-principles pertain to the most pervasive aspects or perspectives of an account; meta-concepts interact with the most definite and restricted elements explicitly described and deployed argumentatively (as concepts) in given metaphysical accounts. All these elements collectively, as a set (although applied one by one), will enable interested metaphysicians to re-examine a position through a variety of abstract approaches about what they have initially posited and concluded within certain argumentative parameters. The results: emendations and refinements improving the explanatory quality of the original visions and articulated understandings and accounts of reality.

From a procedural standpoint, the initial phase of relevance will therefore transpire at the level of heuristic inquiry, that is, discovering whether and the extent to which a given transcendental element can apply with regard to clarifying or critiquing a given metaphysical position. Two important caveats: first, although possible relevance may involve any element among the meta-principles or meta-concepts, not all transcendental elements may, in fact, apply to the statement of factors or the features of reasoning that collectively constitute a given metaphysical position. Thus, the heuristic dimension includes determining whether any of the transcendental elements will apply and, if so, which one (or ones): some of the elements may not. Second, a particular application also includes the possible recognition that what appeared not only initially feasible but perhaps incisive as an application to a given position may not, in the final analysis, represent a viable alternative. Determining, and justifying, such a "negative" result is nonetheless valuable in establishing that some seemingly possible changes or additions may in fact stultify additional investigation given the range of insight and relevance embodied in the scope of the position originally subjected to critical scrutiny. What initially *looks* good as far as making a given position discursively better than it was may not, in fact, *be* good.

To summarize: investigation of a given area of content produced by the application of any transcendental element may yield an opportunity to formulate (a) new concepts (expressible linguistically) available for consideration in subsequent metaphysical situations or to exploit (b) the possibility that existing terminology applies to metaphysical positions in ways that will reveal new aspects of those positions for critical review. Although initially heuristic, an application of a transcendental element (or elements) can indeed become explicitly constitutive and determine a concrete change in a given position but only after (i) integration of the proposed emendation with the conceptual content of the original position and (ii) review of the logical implications generated by the projected additions. A final procedural footnote: The imagination plays an essential role in the process of assigning such a constitutive revision (we will discuss this function of the imagination below).

As a sort of echo of the good as sounded in the *Phaedo*, that is, "holding things together" (contexts in which the good might function in this capacity were discussed in Chapter 1), it is possible to recognize that these sets of basic elements, M-Principles and M-Categories, are held together programmatically but as a *primarily heuristic unity*. An investigator embracing this unity is invited to examine any prior systematic or regional metaphysical exercise according to the specifications of any one, some, or all elements in these sets, and also to consider the implications of such examination. (More specialized strategies for pursuing processes of application are discussed below.)

Continuity and Limits of Inquiry

The transcendental grid represents a section of the divided line. As we argued in Chapter 7, this section is continuous and therefore can be subdivided without limit, thereby capturing an essential aspect of mathematics, that is, that its history has its own underlying structure. Abelian groups represent one example of the mathematical realities generated in conjunction with the transcendental grid. And with the structure of Abelian groups in hand, we proceeded to identify transcendental elements derivable from this type of mathematics examined in relation to one instance of metaphysical inquiry (Chapters 5 and 6). Generally speaking, the parallel functionality of aspects of Aristotle's criticism of Plato's philosophy of mathematics paired with Plato's position on collection and division, in conjunction with the foundational force of the revamped vision of the good (presented in Chapter 7), justifies the procedural inference that some or all of the specified elements of the transcendental grid can apply heuristically to *any* metaphysical configuration.

In conjunction with this conclusion, we restate the principle of equiprimordiality: mathematics and metaphysics share certain structural and conceptual features which allow relevant analyses of mathematics to extract some of these features and to employ them heuristically as modes of assessment directed at particular instances of metaphysical speculation. We also recall that mathematics is inherently historical—a drastic, almost cartoonishly simplified version: arithmetic, geometry, algebra, calculus—and Abelian group theory. But it is vital to keep in mind the argument in the third phase of the reformulation of the divided line, that is, that the continuity of the line as a basic component of the transcendental grid provides a foundation for continual, ongoing mathematical investigations and results. The future of mathematics is extensive, doubtless exponentially vast given the many different areas in which mathematicians have worked, continue to work, and continue to discover novel sectors of mathematical possibility to explore.

In this respect then, mathematical theorizing enjoys a discernibly progressive character; as a result, it may be said that mathematics itself continually becomes "better" than it was. But when a different type of mathematics intersects the transcendental grid, it will doubtless produce a different set of elements—not different in all respects, but different in terms of the specific features of the type of mathematics in question. Any advance in mathematics may therefore provide conceptual material relevant to the structure of a variety of metaphysical positions, material that will illuminate a given instance of metaphysics when these elements have been juxtaposed with the metaphysical position in question. And if the derived elements are indeed applicable to the intersection of mathematics with other metaphysical configurations, then the elements themselves warrant technical consideration with respect to the selection of their names as well as the range of their interpretive functions.

The M-Principles and M-Categories listed above, derived from the examination of *one* type of mathematics in conjunction with *one* metaphysical position, are both introductory and provisional. But now we can maintain that any sets of properties derived from other, new areas in mathematics will become applicable to mathematical analysis in conjunction with the properties called transcendental elements. As future mathematics continues to explore the realities proper to a given mathematical situation, the number (and, we may presume, type) of transcendental elements will also correlatively increase. But these novel dimensions will not affect the fact that sets of new properties will also apply, heuristically, to *any metaphysical configuration*, regardless of the technical, idiomatic style in which a given position may have been originally deployed.

Consider in this regard an observation from Whitehead made a century ago: "Such ages [our own, as of 1925], if they are to avoid mere ignorant oscillation between extremes, must seek truth in its ultimate depths. There can be no vision of this depth of truth apart from a philosophy which takes full account of those ultimate abstractions, whose interconnections it is the business of mathematics to explore."[1]

Whitehead's "ultimate abstractions" can refer to what we have called "transcendental elements." These elements result from intersections of areas in mathematics with the transcendental grid and the resulting generation of properties originating from mathematical considerations; because of their malleable extension, however, they are also open to being applied to metaphysical concepts and situations. The "interconnections" of which Whitehead speaks evoke what a philosophy, that is, in its metaphysical moments, can "explore" when turning its attention to what "the business of mathematics" has produced and will continue to produce into the future. When metaphysics has the imagination and courage to pursue this kind of inquiry, it will indeed "seek truth in its ultimate depths."

The goal is not to "mathematicize" metaphysics; it is, however, to take advantage as fully as possible of whatever dimensions of abstract reality an area of mathematics penetrates and explores so that these dimensions can inform parallel features that characterize metaphysical investigations—that is, what we have identified as transcendental elements. This kind of comparative assessment displays the elevated—but not supreme—status of mathematics and its constituent elements. These elements possess extensive heuristic value, the capacity for producing insight as far as regulating and directing elements of a metaphysical system so that they can be determined according to structural dimensions of a given mathematical configuration. The comprehensive juxtaposition of this configuration with all relevant features of a given metaphysical system or position, insofar as these features are subject to critical evaluation, will maximize their potential to contribute to our understanding of reality.

There are, nonetheless, limits to what mathematics can convey in service to metaphysical concerns. Assuming that properties of the highest level of reality generated by the good can be separated from those realities and then distributed in a modulated way throughout other sectors of reality is one of the speculative foundational points of this book. However, this separation continues to respect the most fundamental metaphysical limits established by the original divided line: in Platonic terms (to state the obvious), Forms are not mathematical realities, and neither Forms nor mathematical realities are material objects. From our present location on the transcendental grid, we discern points of what we might refer to as fundamental metaphysical divergence—if we hew to the dimensions of the reformulated divided line and ascend from where we are on the grid, we will confront a region of realities existing in a purely formal sense; if we descend, material particulars exist, that is, realities that can become subject to mathematical consideration (e.g., being counted as individuals, being included in groups) but which, as such, exist in a realm fundamentally distinct from (a) mathematical realities as such and (b) formal realities, both of which types of reality occupy their own metaphysical realms.

As we "ascend" the mathematical segment of the transcendental grid, mathematics becomes more and more challenging to integrate within a given metaphysical configuration. Movement in this direction heads toward the highest levels of abstraction; in the process, we not only distance ourselves from the pragmatic immediacy of numbered things but also enter reaches of the highest generality—what is often identified as "pure" mathematics—a domain where it becomes increasingly difficult to integrate the mathematical elements characterizing areas of advanced inquiry with any component of reality dealt with in terms of personal or interpersonal dimensions of recognizability.

As we "descend" the transcendental grid, mathematics devolves into a counterpart dimension of "pure" arithmetic. Here numbers become monolithic in metaphysical weight, so to speak, seemingly merging into a virtual identity with whatever entities are counted by the predication of a number (or numbers). When we think of 2 as a number, we tend to associate this thought with two things (Aristotle the realist appreciates and vigorously exploits this tendency, as we have seen). In fact, each entity threatens to become unique ("one of a kind," with the kind in question having only a single instance) and capable of being identified only by using a number as an instrument of reference. Finally, when crossing the area of separation into the realm of reality constituted by material objects, the resulting metaphysical articulations will become equivalent to the direct analysis of material objects as such. Analysis of numbers as they apply to individual, concrete things offers possibilities for metaphysical reflection (indicated in Chapter 1 and its discussions of the theoretical ramifications of Socrates asserting that the number 2 depends on the Form twoness). However, only so much can be said about, for example, the duality of twins *qua* twoness, although much can be said about the modes of reality of two identical (in some ways) and different (in other ways) physical (and interpersonally related) beings.

In sum: The metaphysical conclusions tendered from the arguments in this book concerning the transcendental grid and resident intersections of different areas of mathematics along the grid represent neither pure Platonism (implicitly subscribing to the formal realities residing "above" the transcendental grid) nor pure nominalism + empiricism (magnifying the significance of the material realities dwelling "below" the transcendental grid to the point of limiting reality to only these items). The book argued for, and remains in motion along, a segment of the divided line constituted entirely by mathematical properties, areas of inquiry, theorems, and individual operations—and whatever may be their attendant metaphysical elements and implications.[2]

Strategies of Application

To make a given metaphysical position better (i.e., to move that position closer in terms of greater theoretical force and structural complexity to the limit of explanatory possibility), the metaphysician may take advantage of the commonalities generated by the intersection of Abelian groups with the articulated factors of the transcendental grid. Also, as just argued, intersections of the transcendental grid by other, different types of mathematics will typically introduce an additional range of elements that an interested metaphysician may apply to a given position and thereby, after due effort, make it better in the sense just described.

We offer two large assumptions (in an area of inquiry that could readily be expanded). First, the following strategies presuppose that every serious student of metaphysics knows, has studied, or is acquainted with more than one system, or in the case of a given issue in metaphysics, more than one proposed resolution of that issue. In short, a metaphysician is aware of other options with respect to a given position regarding the analysis of reality. Second, and perhaps more consequential in the long run, "doing" metaphysics is and will always remain to a certain extent a personal exercise. In fact, we may distinguish as a procedural matter between personal and professional metaphysics. Personal metaphysics refers to the ensemble of notions and generalities concerning reality an individual employs, more or less consciously, in order to deal with the world with some measure of success and harmony. Professional metaphysics is what one thinks, says, and writes is "real," that is, a process eventually entering the public domain where the results can be scrutinized and evaluated. Various degrees of disconnect may obtain between these two perspectives on reality—for example, time as it is in itself when philosophically discussed in contradistinction to how we lead our lives in relation to time. Thus, it may happen that a metaphysician working to improve a metaphysical position by applying theoretical protocols (such as those that follow below) may not be entirely successful in separating the realms of the two environments, one entirely lived and the other a complex interplay of what is "felt" to be real with what is "thought professionally" to be real.[3]

The program of application outlined here should be examined with that important proviso in mind. What is offered in this regard is only a suggestion, one way to implement the extensive and potentially illuminating array of transcendental elements. When a metaphysician gets to work and decides to explore the possibilities of making a metaphysical account better, any of the M-principles and M-categories may strike a resonant chord at the outset of inquiry, thereby exciting applications of other elements variously, in a sequence controlled entirely by the collective sensibilities and knowledge of the metaphysician, until the process comes to a halt, whether successful, promising, or barren.

As a prelude to the program of steps embodying the application of the transcendental elements to metaphysical theorizing, we assume the necessity, or perhaps inevitability is a better way of stating the point, that the metaphysician will have another system of reality, more or less thought through, already in place and securely veridical in the metaphysician's mind. If so, then when the metaphysician confronts a competing system or position, fairly standardized modes of appraisal are available. The metaphysician can work from within the stated parameters of a position and: (i) derive contradictions; (ii) raise problems of terminological consistency; (iii) indicate gaps in the account based on evidence from experience as it pertains to the issue; (iv) raise problems based on claims or conclusions from other positions but not conclude immediately (as Aristotle seems to do more than once in his confrontations with Platonic mathematical philosophizing) that a difference between two positions automatically neutralizes the target position and elevates the foundational position above the differing position in terms of credibility and explanatory power.

The ways just described are standard equipment and procedures for the metaphysician. But assume our hypothetical metaphysician intends to implement an approach to metaphysics that will *improve* a given position in the manner described in this book. The following sequence lays out one way to proceed:

1. Determine the articulated limits of the issue that initiated inquiry; then consult another articulated position that extends those limits, even if only in a marginal way.

2. Review the meta-principles and note anything in the area of a "global" reaction, perhaps indicating something crucial or even just incisive, compared with the literal character of what the originating element explicitly says or immediately implies.
3. Juxtapose the meta-categories with the original position to the extent that elements of this position can be deployed according to categories.
4. The key step, or phase, of the process: gather all the resulting data and allow the imagination to "play" with the various intersections, interactions, jostlings, and so on.
5. Note differences. The intersection of similar terms with similar elements as given in the statement of the original positions will facilitate the discernment of differences.
6. Any recognizable difference will become potentially fertile and therefore open to theoretical development.
7. Although the amount of time required for a substantive result pertaining to the position under scrutiny to become evident will doubtless vary from metaphysician to metaphysician, it is also entirely possible that nothing may come to light.
8. But if insight is granted, then it is fair to describe the results as the final phase in a process of theoretical conjunction of (a) basic elements found in the history of metaphysics and interlocking (under the aegis of a revamped approach to the good) as the foundational component of specification and unification with (b) structural features of mathematics, especially when conceived and implemented as a continually ongoing process of investigation and discovery.
9. To the extent that the investigation and its subject matter allow, follow the numerical sequence of transcendental elements (as listed above) and be attentive to any response to any element. This response is, at the moment of initial realization, entirely heuristic, that is, it registers a metaphysician's reaction to a given position without determining whether its content is useful or a mere divergence ultimately resolving itself as a theoretical dead end.

10. If, however, what is heuristic can be beneficial to the original position, the metaphysician is invited to integrate the import of this element with the structure of the original position.

This kind of comparative assessment displays the elevated status of mathematics and its constituent elements, that is, its heuristic value, the capacity for producing insight as far as regulating and directing elements of a metaphysical system so that they—and, by extension, the complete system as a theoretical unity or a particular metaphysical issue—can be maximally determined. The structural dimensions of a given mathematical configuration and the accuracy as well as comprehensive juxtaposition of this configuration with all relevant features of a given metaphysical position insofar as these features are subject to critical evaluation will produce substantial metaphysical content and thus render the original position "better" than it had been.

Herewith a summary of how we have rendered the various sections of the *Republic*'s reconfigured divided line—the transcendental grid, equiprimordiality, constitutive elements generated by the intersection of mathematics and metaphysics, actualization of the transcendental elements applied heuristically to a given instance of metaphysical thought. In a word (or a couple of them), the good at work!

If this approach appears attractive, we should mention what will perhaps be obvious to anyone as a *de facto* limit to this mode of investigation (a consequence noted and briefly discussed in Chapter 7). As the mathematics components of the principle of equiprimordiality become increasingly more sophisticated, the transfers to metaphysical positions become proportionately more difficult to experience and to render into some type of technical metaphysical terminology. The history of mathematics and its advances on many theoretical fronts only renders this circumstance more and more evident. But "more difficult" means more challenging to the abilities of the student of metaphysics to discern what is there to be discerned; it does not mean impossible.

We must appreciate, however, that as areas in mathematics change, develop, diversify, and deepen with respect to their basic axioms, assumptions, and proof structures, these transitions may be viewed collectively as articulated, theoretically unified movements toward the good when viewed as the ultimate transcendence, what the Platonic Socrates claimed was undeniably "there" but inaccessible to human cognition. Mathematics cannot become the Forms, but the metaphysical dimensions of mathematics can, as isolatable factors, approximate the good in terms of perspectives open to that fundamental reality.

If this speculative program of application is granted at least a measure of feasibility, an important procedural consequence follows. It will stand as a mark of prudence to anticipate that as far as strategic suggestions for the implementation of the transcendental grid are concerned, the rate of access to the products of equiprimordiality and their relevance to strictly metaphysical concerns will gradually decrease, at least in terms of the abilities of professional philosophers to establish parallel results in the areas of "pure" metaphysical inquiry. Stated more succinctly, the mathematical components will become more and more abstract and equivalently more difficult to break down into notional components of recognizable metaphysical shape. However, to emphasize the point—decrease means diminish; it does not mean disappear into an empty, theoretically silent ether. Mathematics at its most basic will always be to some degree metaphysical, regardless of the degree of complexity the advances in mathematics will exhibit in its future.

In the present, we have attempted to illustrate the primary thesis of the book by stating a series of implications and applications drawn from Abelian groups. Mathematics has advanced far beyond this level of algebra, of course, but we need not project ourselves into an unknown future in order to illustrate additional possible resources in this area of investigation. We need only to cite calculus. The process aspects of calculus, the unity and resident properties of a function in terms of dependency, independence, number of entities affected by the realities of a function—entire dimensions of reality emerge when calculus intersects with the transcendental grid as did group theory via Abelian groups. Whitehead's call for more process elements in the fundamental elements of mathematics (discussed in Chapter 2) would be addressed, at least in part, by analyzing calculus in order to identify its metaphysical structures and then, following and extending Whitehead's proclamation, by interrelating those structures with appropriate metaphysical concepts. A project worthy of consideration, it seems (see, for example, what Hegel offers in this regard in Remark 2 under "The Infinity of Quantum" in *The Science of Logic*).

The strategies of application sketched here are only outlines. But even in such a foreshortened approach, we must consider one especially important procedural characteristic: applying any of the transcendental elements to a given metaphysical position—even if initially only from a heuristic standpoint—as well as articulating in a publicly available setting the results of such applications will require imagination of high order. The metaphysician's store of experience must be amenable to realizing the possibility of identifying and naming within vocabularies displaying metaphysical character potential insights derived from openness to the strategic processes of application discussed in this section. Is there a way to engage the metaphysician so that the requisite imaginative sensibilities for such labor are pitched to a high level of receptivity and then, ideally, creative productivity?

We might respond, perhaps, that beauty is an especially vibrant reality for engaging the imagination. Even for philosophies of grand scope, however, a consideration of beauty tends to be reserved for its manifestations in nature and the world of art. But although it may appear initially to be counterintuitive, beauty and mathematics have been joined in the thoughts of a number of divergently inspired philosophers (and also, as we shall see, by mathematicians themselves!). Reflection on this unique juxtaposition will make it possible, in due course, to appreciate how imagination functions as a vital factor in the speculative modes of metaphysical analysis analyzed in this book.

Mathematics and Beauty: A Metaphysical Appreciation

According to the glimmers of wisdom supposedly evoked by popular stereotypes, mathematics is hard. But if we listen to certain philosophers, then mathematics, even if hard to do, is also beautiful. Can a type of highly abstract thought that seems to most individuals to be inherently difficult also engender beautiful results? And how does this question pertain to the intricate processes at work in the matter of producing better metaphysics?

A quick review. For Whitehead, the link between mathematics and the good is provided by patterns. But snowflakes display patterns and presumably are the result of natural as well as mechanical processes. In addition, many beautiful works of art are patterned, a characteristic that combines forms of natural technique and the merger of artistic imagination and intention. Whitehead's appeal to pattern, although ostensibly informative, appears indeterminate by virtue of being too inclusive, at least without sustained development as far as illuminating how mathematics, via its seemingly remote and often esoteric patterns, connects to the abstract realities envisioned by metaphysical reflection. If this reservation is fairly stated, what can be said about an apparently even more tenuous connection between mathematical patterns and beauty? And with respect to our current interests, can this connection, if established, contribute to understanding the processes involved in producing better metaphysics? The short answer to this culminating question is yes.

Assistance arrives from a seemingly unlikely source. Earlier, we examined a passage from Aristotle's motto for this book as a sort of frontispiece to help focus reflections on the good in relation to mathematics and its connection, via the good, to metaphysical issues. And in fact, connections between mathematics and rarefied realities such as beauty appeared earlier in Aristotle. Toward the beginning of Book M of the *Metaphysics*, Aristotle maintains that "those who assert that the mathematical sciences say nothing of the beautiful or the good are in error. For these sciences say and prove a great deal about them." He continues: the chief forms of beauty are "order and symmetry and definiteness, which the mathematical sciences demonstrate in a special degree" ($1077^{a}33$-$78^{b}1$) Furthermore, the passage in Book M of beauty's scope with respect to mathematics, although brief (and quite apart from whether or not we agree that beauty itself can be explicated in this way), surely adds a measure of theoretical heft to Whitehead's spare references to pattern, since it seems helpful to construe the patterns of mathematics according to how the more specific characteristics of order, symmetry and definiteness cohere with an understanding of beauty.

Further inferences of Aristotelian character are possible, and usefully so, in a related context. We will recall that in his motto, Aristotle spoke of the serious nature of determining how "the elements and principles" of mathematics are related to "the good and the beautiful." Compounding the difficulty of this question is the problem "whether any of the elements is the sort of thing we mean in speaking of the original good and the best, or whether this is not so, these being produced later" (1091^{a}32-33). In the earlier passage (from Book M), Aristotle asserts (1078^{a}32) that the good and the beautiful differ—the latter "always implies conduct as its subject," whereas the former "is also found in motionless things." But if we continue to concentrate on good in its primary metaphysical sense, then the passage from Book N just cited may contribute to the Aristotelian position on connections between beauty and mathematics.

Aristotle doubts whether any of the "elements" advanced by Plato and the Platonists (and subjected to sustained criticism in *Metaphysics* Books M and N) are satisfactory if we speak of "the original good and the best." The passage at 1078^{b}1—appealing to beauty in terms of order, symmetry and definiteness—perhaps then renders juxtapositions such as that between the good, taken in the foundational sense, and the beautiful somewhat less enigmatic and more accessible to understanding. This specification of perspectives on beauty also establishes platforms for additional inquiry and consequent insight into the sectors of reality opened up and explored, even if only to a provisional degree, by initial efforts at speculative investigation concerning these perspectives. A suggestion: this book's derivations from the good as it informs the divided line—that is, equiprimordiality and the transcendental grid—may be taken as an example of what Aristotle maintains could be "produced later," assuming that the slate of assumptions and arguments introduced and discussed in Chapters 7 and 8 be allowed as legitimate sectors of inquiry (and setting aside the possibility that an Aristotelian may have a less than sanguine reaction to the conclusions drawn in Chapters 7 and 8!).

This line of interpretation sanctions an important clarification of the place of beauty in our discussion. What is beautiful in mathematics is not elements or phases of mathematical processes or results proper; rather, the beautiful dimension of mathematics refers to the transcendental elements animating the structures of mathematics *and* metaphysics. To speak of the beauty of mathematics is therefore to incorporate perspectives on those sectors of reality that mathematics shares with metaphysical investigation. And yet, useful though this line of thought may appear, its relevance to beauty itself as well as to the import of beauty with respect to the operations of the imagination and the matter of pursuing "better" metaphysics still remains in a realm of theorized potency, conveying a lingering sense that more could and indeed should be said.

Another source, again seemingly of inapposite provenance, provides additional clarification of the proposed connections between mathematics and beauty. In an essay, "The Study of Mathematics," originally published in 1907, Bertrand Russell (1872–1970), who when the literary mood struck would not hesitate to let his acidic wit inflect virtually any context, adopts a lyrical and almost rhapsodic tone in his account of the ways in which mathematics displays beauty:

> Mathematics, rightly viewed, possesses not only truth, but supreme beauty—a beauty cold and austere, like that of sculpture, without appeal to any part of our weaker nature, without the gorgeous trappings of painting or music, yet sublimely pure, and capable of a stern perfection such as only the greatest art can show. The true spirit of delight, the exaltation, the sense of being more than man, which is the touchstone of the highest excellence, is to be found in mathematics as surely as in poetry. What is best in mathematics deserves not merely to be learnt as a task, but to be assimilated as a part of daily thought, and brought again and again before the mind with ever-renewed encouragement.[4]

This affective passage may well be attractive to those receptive to reflecting on a juxtaposition of beauty and mathematics; but that said, and given the context of the present inquiry, the extensively imagistic dimension displayed in the passage also requires expansion and clarification. What, for example, could Russell have in mind by the appeal to "stern perfection" exemplified in mathematics, especially in tandem with the same quality and range of experience found only in "the greatest art"?

Consider as a possible response to this question the rigorous rules of inference controlling the production of a proof in symbolic logic. To absorb the proof as a whole, a completed demonstration, is to win a sense, subtle but real, of the interconnection of the entities referred to by the various steps in the proof. If the number of steps is the minimum required to establish validity, then this exercise has engendered an instance of a kind of perfection insofar as the process of proof follows relevant rules, and only relevant rules. By so doing, the proof sternly and inexorably proceeds toward its goal without (a) any argumentative movement that might lead in a direction other than that of the intended destination or (b) introducing steps that may cohere with the point of the proof but are more numerous than necessary for the proof to reach the desired conclusion. The proof's written steps, moving in immediate and direct sequence, register on the mind and excite a rarified but nonetheless palpable and, surely, attractive awareness of order, necessity, and power.

We offer this set of circumstances as one possible explication of "stern perfection"—here is another. Russell mentions music in his ode to mathematical beauty; if we felt more comfortable with a sense of beauty occurring in a setting defined by more immediately accessible modes of experience, then the beauty of mathematics might approximate what a brief (or, indeed, extended) session with the aural grandeur a master keyboard player can produce while working through sections of J. S. Bach's "Well-Tempered Clavier," music which, as such, is just another configured set of notes on a score (i.e., sheets of paper)—until the notes are played by someone who knows and feels how best to play them. Although the differences between a "beautiful" position in mathematical demonstration and a beautiful performance of Bach's masterpiece are many, the dimensions of delight characterizing both experiences reflect the theoretical rightness in Aristotle's instinctive yet nonetheless broadly diversified recognition of beauty in mathematics, as well as Russell's more nuanced expression of the stern perfection structuring the same basic revelatory reality.

Russell's account of the relation between beauty and mathematics also refers to what is "best in mathematics," but, viewed again from the perspective of interpretive clarity, it does not illustrate how to get a feel for, or any sort of cognitive control of, this sense of "best." Are some types of mathematics intrinsically "better" than other types, suggesting the possibility that a certain type is "best"? Or is the intended point less speculatively evocative, that is, that regardless of what area of investigation is under scrutiny, some treatments of certain mathematical material are "better" than other treatments of the same material (which would sanction nominating one such treatment as the "best" of those available on a given issue)?

A suggestion derived from the direction of argument pursued in this book: Russell's appeal to "best" may be understood within the boundaries of how a given metaphysical position could be made "better" by applying the procedures established in accordance with the relevant mathematical features of the transcendental grid. Thus, what is "assimilated" in terms of Russell's claim could be restricted to what would come "before the mind" of the metaphysician if this individual were to reflect on reality as "a part of daily thought." As a result, the metaphysician's account of reality would become more comprehensive, more detailed in terms of property designations, more explicit in terms of possible relations between a given entity and what exists other than that entity, and more present to ordinary modes of perception, less esoteric and theoretically remote. In a word, the assimilated account would be beautiful by virtue of being metaphysically more complete.

Another advantage of this interpretive turn is that these suggested ramifications in theory, as derived from Russell's evocative prose, also speak to Aristotle's undeveloped survey of the beauty of mathematics in terms of "order, symmetry and definiteness":

Order—by the increase of ways in which a given entity establishes connections between and among any novel additions to a given metaphysical position regarding the abstract structure of that entity;

Symmetry—what emerges when a metaphysical element is extracted from a mathematical position and then integrated within the fabric of a target instance of metaphysics, that is, the coincidence and sharing of a common characteristic (from a musical standpoint, a medley of samenesses and differences), the evenness of balance and harmony recognized as patterns governing the interaction of distinct entities and regions of entities;

Definiteness—the characteristic of informed and illuminating clarity brought to the metaphysical target upon the integration of the novel element, with the resulting sharp specificity in the description of each entity insofar as it retains its autonomy as this or that thing but displays this autonomy in intimately articulated relations to other things within the uniform presentation of the metaphysical account under scrutiny.[5]

If the above summary fits our context of discussion, then from a collective point of view, we have deployed the Aristotelian approach to beauty as a mathematical reality in terms marked by a high degree of abstraction; this is not beauty understood as the intense realization of a sudden gush of feeling when one is touched by some sort of tangible (and, in the case of music, fleeting) phenomenon. Yet the attractiveness of the effects of mathematics when experienced in the compressed metaphysical contexts Aristotle indicated is nonetheless real and significant to appreciating the proposed connection between beauty and mathematics.

It is not then an interpretive exaggeration to view Bertrand Russell's lyrical approach to this connection as redolent with possibilities; as such, what Russell tells us stands as a worthy foil to parallel passages in Aristotle (which we have just explored). However, to be in an even better position to realize and articulate what is said toward the end of the passage quoted from "The Study of Mathematics"—which can be seen as "beautifully" tangent to the central role Abelian groups have played in this book—it will be useful to broaden the theme of our inquiry for the purposes of increasing access to the complexities and subtleties of beauty when viewed from a mathematical perspective. To accomplish this end, we will also amplify the ensemble of voices addressing this more extended philosophical landscape. The philosophers to follow in the next section may be viewed as a small chorus intoning their thoughts on preliminary matters leading to the celebration of the crucial importance of the imagination. Note: Russell will return toward the conclusion of the next section.

Mathematics, Metaphysics, Imagination

In "Of Liberty and Necessity" (§8 of *An Enquiry Concerning Human Understanding*), David Hume explains how human beings arrive at the notion of necessity, an issue relevant to the possibility and meaning of freedom. Hume posits an epistemic necessity that derives from the sameness in basic cognitive and emotive experiences, that is, the fundamental character of acknowledging internal and external impressions with the consequent formation of ideas based on these impressions—a core position in Hume's epistemology (and developed in Sections 2 and 3 of the *Enquiry*). This kind of necessity then produces a crucial result in terms of human behavior and the understanding of such activity:

> It is universally acknowledged, that there is a great uniformity among the actions of men, in all nations and ages, and that human nature remains still the same, in its principles and operations. The same motives always produce

> the same actions: The same events follow from the same causes. Ambition, avarice, self-love, vanity, friendship, generosity, public spirit; these passions, mixed in various degrees and distributed through society, have been, from the beginning of the world, and still are, the source of all the actions and enterprises which have ever been observed among mankind.[6]

Hume then draws a conclusion (55) with, as we shall see, implications for the possibility of engaging the imagination for purposes of producing a "better" metaphysics: "Would you know the sentiments, inclinations, and course of life of the GREEKS [*sic*] and ROMANS? Study well the temper and actions of the FRENCH and ENGLISH."

The first point concerning these remarkable passages is to appreciate the fact that Hume, naturally suspicious of general—or, as he sometimes disparagingly says—"metaphysical" ideas, seems committed to a kind of universal identity regarding the sameness of "human nature" (cf. Plato or Aristotle) at least in terms of motivations for why we act as we do, and the results of actions based on such motivations. But more to the issue at hand—Hume claims that studying "the temper and actions" of the French and English would enlighten us concerning Greeks and Romans (presumably the ancient forerunners of eighteenth-century Greeks and Romans). If we want to know more about the ancient Greeks and ancient Romans, then Hume's suggestion is to study the contemporary French and contemporary English. Since, for Hume, the basic experiences stipulated in his list are the same for all, then whatever holds for the French and English we see around us also held for Greeks and Romans who lived millennia ago.

But consider for a moment just the French and English (apart from their earlier Greek and Roman counterparts)—if our interest in comparative exponents of humanity were limited to the motivations that produced philosophy, in particular metaphysics and what Hume identified as "human nature," what would Hume say if we juxtaposed Jean-Paul Sartre's *Being and Nothingness* (1943) and Gilbert Ryle's *The Concept of Mind* (1948)? Would we say that the types of human experience that funded each of these works as well as the assumptions and commitments about reality permeating both of them were as uniform and constant as Hume maintained? Consider another such pairing: F. H. Bradley's *Appearance and Reality* (1893) with Friedrich Nietzsche's *Beyond Good and Evil* (1886—in the Preface to which Nietzsche asserts that the "worst, most durable, and most dangerous of errors so far" was "Plato's invention of the pure spirit and the good as such").[7]

Hume could not have anticipated the specific works cited in the above juxtapositions, but this fact about the content of future philosophizing is by the way; after all, Hume said with great confidence that human nature always remains the same "in *all nations and ages*" [italics mine]. The historical and comprehensive universality of the claim is undeniable. Nevertheless, the suspicion may arise that Hume's appeal to a common storehouse of experiences and human characteristics, allowing the interested student of humanity to move freely between and among any past or current forms of human expression and draw reliable conclusions about individuals from different nations and cultural backgrounds, seems provincial, if not perhaps distressingly naive. And yet, at some levels of investigation, reality for all of us is, surely, just what it is. All human beings are warmed by the same sun, all have material bodies, all eat food, and all interact with particular animate and inanimate things (in addition to, for most of us anyway, other human beings) which are in some sense all "out there." If we accept Hume's position on the complete uniformity of basic human experiences and motivations, then why have those philosophers authoring treatments devoted to stating what is and is not real issued such vastly different generalized versions of what we all share and act with and through? In a more restricted context of inquiry relevant to our immediate concerns, what motivates and enables metaphysicians to pursue the kind of theoretical work they do, speculative thought seemingly so distant from the routine and immediacy of "standard" or "typical" human affairs? Why have philosophers, in particular those concerned with metaphysics—that is, veridical and justified descriptions of reality—produced such vastly different accounts? Is there any human medium through which we might gain at least partial access to Hume's layering of bedrock circumstances and experiences shared by all?

This question may not be answerable, especially if a projected reply were proffered in the face of the fundamental epistemic samenesses Hume has identified. But the fact of the matter is that differences of vast dimension mark the range of the multivarious texts the metaphysician must admit as seemingly suitable to the metaphysical enterprise and thus as at least potentially viable voices to allow entry into the discussion. As a result, whatever may be the explanation for why the Western tradition's metaphysicians have produced such vastly different accounts of reality, if in the final analysis there is only one reality which we all share—and interact with—in basically the same ways, a metaphysician who at least countenances the potential relevance of divergent accounts of what is, at its core, the same subject matter for everyone confronts a serious and labor-intensive decision: some effort must surely be made to acknowledge, if not to attempt to integrate, relevant features of other metaphysicians' work into whatever a given metaphysician has initially produced. If so, how then to proceed?

Consider three thinkers—Descartes, Leibniz, and Whitehead—in observations concerning the activity and limits of the *imagination*, an element of humanity (or, perhaps, with a subtle bow to Hume, "human nature") in which we will assume all of us share, at least in some respect.[8] The initial juxtaposition and discussion of Descartes and Leibniz are suggestive for two reasons: (1) both philosophers were also epochal mathematicians—Descartes with analytic geometry, Leibniz with calculus; (2) both philosophers introduced the imagination at crucial points in their metaphysics. After brief discussions of Descartes and Leibniz, we then return to the mathematician and philosopher Whitehead on the imagination when viewed, again, from the standpoint of the relevance of imagination for the metaphysical enterprise, especially when Whitehead's position is inflected with the factor of beauty.

Toward the beginning of Meditation 6 in the *Meditations on First Philosophy* (1641), Descartes is exploring possible avenues of thinking in order to prove the existence of material things. He begins by considering the imagination, which earlier, in Meditation 2, had been identified as one function of thinking. In Meditation 6, imagination is defined as "simply a certain application of the knowing faculty to a body intimately present to it, and which therefore exists."[9] This definition implies that if bodies did not exist—which, at this point in the *Meditations,* remains a theoretical possibility—then imagination would have nothing on which to work (thereby suggesting in turn that imagination itself would not exist since it lacked a necessary and palpable reality in order to function in any respect).

In fact, Descartes pursues another line of thought. He argues that a chiliagon, a thousand-sided figure, requires a "peculiar sort of effort on the part of the mind to imagine [compared to, say, a pentagon], one that I do not employ in order to understand" (48). Descartes reasons that imagination and pure intellection are thus fundamentally different. If so, what follows? Descartes forthrightly asserts that the dependency of a functioning imagination on a (problematically existing) body external to it justifies, at this point in the overall argument of the *Meditations*, the conclusion that the imagination could be jettisoned entirely from the ensemble of functions characterizing the thinking thing: "For were I lacking this power, I would nevertheless undoubtedly remain the same entity I am now" (48). For Descartes, I would remain a thinking thing in all relevant respects even if I, as a thinking thing, lacked the imagination.

Not many years later, Leibniz contends (in Section 6 of the *Discourse on Metaphysics*, 1686), that nothing God does is not orderly; thus, everything existing is in conformity with the universal order. "This is true to such an extent that not only does nothing completely irregular occur in the world, but we would not even be able to imagine such a thing."[10] Leibniz imposes a restriction on the imagination with respect to the possibility of holding in the mind an image of sheer disordered randomness. The imagination is structured in such a way that it simply cannot process such an image. The claim is not that such randomness is inherently contradictory, for then the argument could be that we cannot imagine anything, regardless of content, which displays the logically determined set of circumstances that define contradiction. Rather, the point seems to be that the utter alien strangeness of randomness, especially when situated within a universe divinely created—and ordered—in all respects, obviates conjuring up, in any imagistically convoluted form or pattern, even the possibility of such randomness.

If we review these two positions, it seems fair to conclude, in summary fashion, that for both Descartes and Leibniz, the imagination is an inherently limited feature of the human animal. How then should the student of metaphysics pursue this area of philosophy if, at some juncture in a thoughtful response to reality, the student feels it necessary to present something new for the consideration of other students of the same subject? From what inner source could language and concepts originate that would accommodate the felt, novel visions of this metaphysician?

In "Speculative Philosophy," the first chapter of *Process and Reality*, Whitehead asserts (4) that in philosophy, "deficiency in imaginative penetration forbids progress in any form other than that of an asymptotic approach to a scheme of principles only definable in terms of the ideal which they should satisfy." If philosophy at its most vibrant is speculative, this deficiency results in philosophical activity which, if pursued as metaphysics, at best moves toward a restatement and marginal development of whatever lies as "ideal," that is, existing only as potentially realized (or realizable) as prescribed by a given scheme of principles. Such material cannot display more than an ossified, indeed programmatic staleness ultimately leading to a lack of relevance to what, for Whitehead, is the ongoing and continually diverse and also captivating drama of human experience and cosmic processes. Given this approach, it is obvious that the ways in which he envisioned how imagination functions are dramatically different from what we read in Descartes and Leibniz. For Whitehead (5), the true method of philosophical discovery is like a plane in flight: "It starts from the ground of particular observation; it makes a flight in the thin air of imaginative generalization; and it again lands for renewed observation rendered acute by rational interpretation. . . . It can even play with inconsistency, and can thus throw light on the consistent, and persistent, elements in experience by comparison with what in imagination is inconsistent with them."

A remarkable and, perhaps, inspiring image, expressed with considerable feeling. Whitehead appreciated that if reality was in fact as complex in its processes as he credits it to be, our cognitive apparatus must be equipped with a distinctive capacity for absorbing as much experienced data as possible in order to be in a position to coordinate those processes within the ambit of theory as fully and accurately as possible. With hindsight (i.e., the discussion in Chapter 2), we may observe how Whitehead employed imagination in connecting the Platonic use of triangles in the *Timaeus* with modern scientific notions of molecular and atomic structure as they inform realities in the natural order. Indeed, the present book can itself be seen, at least at certain junctures, as an approximation of Whitehead's visionary procedural perspective. If, for example, we take the "thin air of imaginative generalization" to refer to the entire process of speculative possibility regarding metaphysical reflection, then a relevant example could be the theoretical locale in which appeared the substantive results of inquiry laid out in Chapter 7, in particular the notions of equiprimordiality and transcendental grid. These notions (as discussed earlier in Chapter 8) then landed for "renewed observation rendered acute by rational interpretation," that is, the relevance and potential fertility of human awareness when actively in the presence of the transcendental grid when a type of mathematics—for example, Abelian groups—intersects the grid.

In the case at hand then, the imagination houses the capacity for the metaphysician to recognize and articulate the reasoned statement of the limits of inquiry based on the differences between formal realities and material objects residing at opposite ends of the transcendental grid. As a result of this experience, the metaphysician relocates the operative sensibilities to reality and will be in a position to state a renewed observation of consequences rendering a given instance of metaphysics "better" than it was by means of reasoned application of commonalities in mathematics and metaphysics—that is, the transcendental elements—with the results of such reasoning integrated within the stated structure of the original venture in metaphysics.

If Whitehead is correct, the results of the discussions in Chapters 7 and 8 provide examples of the imagination at work, in his phrasing, "imaginative penetration" of the Platonic notion of the good and the suggested reformation of that notion. Armed with this revitalized account of the existence and importance of imagination, we may now attempt to clarify the conclusion of Russell's description of the beauty of mathematics. Here is the relevant section of the quoted passage: "What is best in mathematics deserves not merely to be learnt as a task, but to be assimilated as a part of daily thought, and brought again and again before the mind with ever-renewed encouragement." Russell's "assimilated," and the phrase "ever-renewed encouragement," are replete with process dimensions. It becomes a question, therefore, what could be meant given that Russell does not identify the target audience for these recommendations as professional mathematicians or even philosophers, but simply and directly "to the mind," presumably to any and all human beings whose minds are sufficiently receptive to absorbing the relevant aspects of "what is best in mathematics." Can we clarify how these processes should be understood, especially regarding their relevance to the reality of beauty?

As an illuminating supplement to Russell on the beauty of mathematics, consider this comment by Whitehead from *Science and the Modern World.* The context is the progress of mathematics in the seventeenth century, but the point Whitehead makes has far more pervasive implications, especially for the issue under examination: "Nothing is more impressive than the fact that as mathematics withdrew increasingly into the upper regions of ever greater extremes of abstract thought, it returned back to earth with a corresponding growth of importance for the analysis of concrete fact" (32). If we juxtapose the passages from Russell and Whitehead, an intriguing parallel emerges—Russell moves from what is best in mathematics to a plea for its integration into daily rounds of pragmatic awareness of anyone capable of appreciating this level of reality; Whitehead sees a connection between "the upper regions of ever greater extremes of abstract thought" and the importance of such reflection "for the analysis of concrete fact." The parallel is from "higher" mathematics to "lower" and particularized sectors of reality. Although Whitehead's point is not made in the context of connections between beauty and mathematics, his emphasis on the relevance of mathematics—indeed, as it ascends into its "upper regions"—for concrete fact brings out how mathematics can show us important (and perhaps hidden) aspects of this or that particular set of given circumstances as they become embodied in "concrete fact." If we shift this revelatory activity to Russell's frame of reference, then we would need to show the senses in which Whitehead's appeal to a certain type of concreteness plays into what Russell wants to be "assimilated" within the daily fabric of life, where particularity and practicality dominate as the typically unspoken watchwords controlling human affairs.

It is, perhaps, the democratic flavor in Russell's assertion that is so striking when we attempt to integrate, with continuous encouragement, the lessons initially realized in the minds of mathematicians into the daily conversation and experience of people in general. For this very large audience, particular things exist autonomously as instances of unity, as this or that thing. But they also exist in concert with many other things of the same and different types. Thus, they exist in groups, sometimes small in number of elements, sometimes with many other elements. Group theory in mathematics recognizes this feature of individuality and unity, deploying it with respect to considerations that allow the unity of an individual thing to be appreciated for itself and also in relation to other things (and, in turn, the exhibitions of the unities of these other things).

From this perspective, beauty becomes a complex property that provides extensive specification of its attractiveness on occasions when beauty is predicated of a particular thing viewed under the aegis of a mathematical group. This approach yields insight into the unity and uniqueness of particulars through the presence of the kinds of theoretical discrimination provided by the technical definition of an Abelian group and recognition of the realized transcendental elements when that area of mathematics intersects with the transcendental grid. If indeed it is possible, with Russell, to have human beings regularly educated in ways that enable a properly imaginative assimilation of such medleys of unity and difference, then even minds not especially gifted with mathematical ability proper will be in a position to experience particulars in ways that will result in these individuals' appropriate use of words such as "beauty" and "beautiful," as well as their occupying, even if for only a few moments, the levels of interior appreciation and excitement duly inspired by such words.

As Russell perceptively observes, this kind of rarified education requires "ever-renewed encouragement." No single thing should occupy the attention of anyone, not when reality is populated by so many other things of the same, similar, or different stripe. The controlling structure of a mathematical group, when "assimilated" (a fine choice of words here!) within the fabric of common experience significantly contributes to the recognition and experience of this level of beauty. Minds so educated, especially from the standpoint of enhancing and enlivening the imagination, will be in a position to recognize and duly appreciate beauty in the many instances of particularity open to human experience. These minds will also be as receptive as possible to the attractiveness of generalizations when, in mathematical and metaphysical settings, generalizations encompass the specific and individual particulars of "common life" (a favorite noun phrase in Hume's *Enquiry*) and exhibit a multitude of variant connections between and among them. But, and the point surely bears repeating, given the popular resistance to an activity as "hard" as mathematics, these subtle but real connections between mathematics and beauty, via the organizational depth and intricate structure of group theory, must become an oft-repeated process to the intended audience in order to achieve a level of ready accessibility even when the imagination has been primed to become as receptive as possible to the recognition of these connections.

Imagination Revisited

The previous section included brief reviews of Descartes and Leibniz as well as a more extended treatment of Whitehead, all on aspects of the imagination. Viewed collectively, the three accounts illustrated the disparities among major philosophers on this subject, on what imagination is not, and is, capable of realizing within human consciousness. It should be informative then to offer a summary account of the primary functions of the imagination with respect to its contributions to the metaphysical enterprise, given what has been asserted on this theme in the previous two sections. We will focus on a transition from (a) approaches to understanding the beauty of mathematics to (b) how this sense of beauty serves as an initiating, transitional reality through which we can understand the implied structure of the imagination as it has taken shape throughout the current analyses.

We have sketched an interpretation of Russell's desire to make mathematics come to life in the practical matters of human experience, the typically mundane web of what people do as a matter of course. This plea suggests that the parallel elements of metaphysics, via the principle of equiprimordiality, may offer a realistic "live" substitute for the sheer abstract remoteness of mathematical concepts and its modes of demonstration. The counterpart experience on the part of the practicing metaphysician therefore becomes recognizing and appreciating the inclusion of a given item of reality which, prior to the reconfiguration of a given position, had been unnoticed and unanalyzed—and therefore, in a very real sense, non-existent from the standpoint of the dimensions of reality appearing in an explicitly stated metaphysical position.

With due acknowledgment of Russell's egalitarian approach to the reality of beauty, we may note then that every human being is a potential metaphysician by reason of existing in relation to self and the "outside" world in varyingly aware ways. During such receptive modes of awareness, an individual might register the existence in consciousness of something, an aspect of a thing, relation of a thing to something else, that is, which had not been noticed by the working metaphysician to the extent of that phenomenon becoming explicitly integrated within the theoretical confines of a given metaphysical account. The challenge becomes to access this experience, review and analyze its contents, note the incipient sense of a structure, make its presence felt as fully as possible within one's awareness, and then to express it in some mode of language (including the transformational languages of artistic media) so that the experience as well as the underlying reality can be noticed and, perhaps, shared by others. Would the metaphysician be up to a challenge of such dimension? Regardless of any *de facto* realizations of this possibility, it seems fair to maintain that the more avenues of accessibility we make available for scrutiny—including, as this book has argued, the complex and diversified world of mathematics—the better opportunities we command to increase our articulated awareness of reality.

We conclude this section with reflections on an especially intriguing passage from Whitehead's *Adventures of Ideas* concerning the relation between metaphysics and imagination. Implications from this passage provide avenues of access to the imagination itself, especially from the standpoint of the imagination viewed as productive in nature. The discussion also stands as a brief but worthy penultimate statement on the primary point of this book.

The passage in question is from the chapter "Cosmologies," and the context is a discussion of rival approaches to understanding how best to construe the cosmos when examined at a fundamental level of abstract inquiry: "Our metaphysical knowledge is slight, superficial, incomplete. Thus errors creep in. But, such as it is, metaphysical understanding guides imagination and justifies purpose. Apart from metaphysical presupposition there can be no civilization."[11]

If Whitehead is correct in his assessment of what we know metaphysically, then it seems obvious that we should do whatever we can in order to rectify, to whatever degree of improvement is feasible, the "slight, superficial, incomplete" quality of our metaphysical knowledge. The collective recommendations discussed in Chapters 7 and 8 of this book may provide a potentially suitable program to consider in this regard. However, Whitehead's reference to the fact that "metaphysical understanding guides imagination" seems to add another dimension to the process of applying the transcendental elements to metaphysical accounts for the purpose of making them better. Our discussion so far has emphasized the imagination working on its own, as it were. But Whitehead tells us that, in fact, it is "metaphysical understanding" which, when accompanied by necessary technical elements such as definitions and concepts, becomes capable of guiding imagination to accomplish what the province of the imagination has the capacity to produce in these areas of inquiry. What can be said about the imagination as such then, assuming that its operations have been inspired and, to some extent, controlled by this kind of abstract understanding?

Our approach has been to have "best" (understood from the fundamental perspective of the good in Plato) yield to "better" (as in a better account of a metaphysical position), and mathematics in a pure sense being relocated via equiprimordiality to instances of realized transcendental elements as they emerged from the transcendental grid. We then discussed the inspirational, triggering effects of beauty on the imagination—the "austere" beauty of which Bertrand Russell speaks. Furthermore, we interpreted features derived from the structure of Abelian groups, via the transcendental elements, as capable of evincing beauty within the demands and enjoyments of daily life, thereby filling in the gaps, as it were, left over from a glut of abstraction while pursuing the theoretical demands of classically oriented metaphysical thought.

A brief pause is perhaps in order. Given its complex and subtle subject matter—mathematics and its correlations with metaphysics as just described—it would seem that Whitehead's "thin air of imaginative generalization" may be in some respects ironically identified. A more appropriate image than "thin air" for the imagination's function as Whitehead envisions it is a sort of crucible in which whatever has been deposited from routine experience becomes, or at least can become, energized to an extent allowing something significantly different in terms of its metaphysical description to emerge. In fact, a more theoretically grounded source for such significant emendations on the fabric of experience would be something like the Kantian notion of the productive imagination (as introduced and developed in Section 49 of the *Critique of Judgment*). For Kant, the productive imagination is that capacity of the inner self where a "second nature" is created based on the material provided by the appearances of actual nature.

Kant's systemic interest at this point in the third *Critique* is the production of art, but given what Whitehead wants the imagination to accomplish (and how we have interpreted its function here), it would be legitimate to attribute to the productive imagination the capacity to generate a much broader and, in some respects more robust range of "secondary" realities. Thus, at the level of practical immediacy, the productive imagination is, for example, where undergraduates can glimpse their future occupations (and then decide how much philosophy to take) and, in a setting closer to its stated purpose in Kant, the productive imagination is where artists visualize a tangible item in their favorite medium that will help pay the rent (or, perhaps, one day rank as a masterpiece). But at the other extreme from such immediacy, the productive imagination would also evoke an environment organized with materials for rectifying any weaknesses (if not errors) as well as enhancing the explanatory effects of a metaphysical discussion by recognizing what would be required to make that discussion better than it was. We may construe this to be a signal activity of the productive imagination, assuming that this capacity were scrutinized in terms of the project pursued in this book.

The notion of the productive imagination in fact offers a host of possibilities in this area of metaphysical exploration. As a further illustration of its fertility, consider the fact that the imagination is employed in a variety of dimensions and plays ongoing yet varying roles throughout the processes involved in making metaphysics better. Initially and broadly stated, the imagination provides a global realization that elements of mathematical entities, if approached from a certain conceptual standpoint, can function at a level apposite to elements of certain metaphysical positions. Imagination produces an awareness that these commonalities exist, and that when analyzed with respect to a given position, the commonalities may generate consequences relevant to the improvement of structural considerations pertaining to the original metaphysical position. If this process is examined in closer detail, we can discriminate a series of stages showing how the imagination integrates itself within the process phases that arise when instances of transcendental elements are realized along the transcendental grid:

A. Recognizing the appropriate transcendental elements (indeed, perhaps all of them!) to juxtapose with the features of a given metaphysical position: We understand the direction of argument from the principles and premises explicitly asserted in a metaphysical position; however, it is imagination that approaches the stated boundaries of these positions as potential points of departure, alive and in motion, rather than as ultimate and fixed theoretical destinations.

B. Determining whether the juxtaposition of a transcendental element and an aspect of a target metaphysical account is, indeed, *not* revelatory in some respect (the imagination becomes a key instrument in the heuristic dimension of applied transcendental elements). When this kind of investigative dead end occurs, Whitehead's alert concerning the directive force of metaphysics comes into play and, as a result, the student of reality moves on to other matters (which might include the juxtaposition of a stated factor in the original account with a different transcendental element).

C. If the juxtaposition looks potentially revelatory, imagination provides an initial feeling of structure as well as the theoretical shape of that structure.

An extended example: Descartes' account of the workings of the imagination began with the immediately present image of a body—and, unfortunately, ended there. This initial identification is nonetheless on the right track since imagination must start its work by paying attention to something other than itself. But an operational productive imagination would have poked around, so to speak, on the other side of the uniform material "covering" of the body as this view is immediately given to perception and, in this cognitive format, presented to the imagination. The attraction of this image as a single, unified phenomenon would have excited the imagination to move into position so that relevant functions of human understanding could provide reasons accounting for this attraction. Why do we think and act as if this image is of just *one* body? Imagination would not stop there, in the midst of exploring possibilities with respect to an imaged unity. The exploration would continue so that those reasons would be couched as often as possible within the structure of this unity to the extent that this structure is deployable in vocabularies of general, metaphysical elements. What must be the case—and how would we, with the assistance of the originating power of the imagination, say what must be the case—in order to allow for the possibility that one material thing can exhibit this particular image of itself?

D. Imagination presents a rudimentary survey of the perspectives in which existing language can be reformulated to represent that feeling of structure; if nothing arises from existing language and the felt sense of structure persists, then it is a crucial function of imagination to conceive and linguistically identify, or perhaps even coin, a bearer of new and incisive metaphysical significance.

The accounts of the imagination and how it works, as presented in this and the previous section of Chapter 8, must suffice as far as offering perspectives on its function as a phase in the cognitive activity required to implement programmatic suggestions for making a metaphysical account better than it was.[12] The continual activity of practicing mathematicians worldwide certainly seems to justify the assumption that mathematicians can take care of themselves regarding coordinating the capacities of imagination required to master and extend their discipline, and to display in public and with due rigor its many dimensions of austere beauty. But what can the philosopher, especially the metaphysician, hope to take away from such imaginative recastings and juxtapositions?

Our response takes the form of a brief restatement of the book's main argument: Plato on number and philosophical method, modern abstract algebra and Abelian groups—this book has explored some of what can be said when these expressions of philosophy and mathematics meet and mingle. We argued that a detailed and conceptually developed approach to the good—modeled after the Platonic notion of the good but grounded in specific metaphysical elements—is directly applicable to advances in modern mathematics (e.g., group theory). What we identified as transcendental elements, derived from this mathematical foundation and filtered through an imaginative prism, can then be used heuristically as a comprehensive critical schematic for assessing and potentially expanding any extant metaphysical position or system.

We recall, by way of illustration, that the arguments and discussions in Chapters 5 and 6 provided more detail in an actual context of philosophical criticism (Aristotle on Platonic philosophy of mathematics) and application (juxtaposition of Aristotelian criticisms on Platonic metaphysics, that is, collection and division). Also, when even more sophisticated areas of mathematics intersect with the transcendental grid, additional transcendental elements may become available. The interested metaphysician, especially if endowed with an appropriately productive imagination, then has an even greater opportunity to analyze the original formulation of an issue (or a complete system) and to make that account better than it was prior to this appositional exchange with the metaphysical elements that underlie and drive any mathematical system of thought. As a result, to imagine producing a metaphysical account displaying these characteristics would represent, in its own way, a beautiful possibility.

Pursuing Reality: Metaphysics and a Sense of Adventure

A final, perhaps provocative thought on some personal aspects of "doing metaphysics" that may not be obvious even to those in philosophy who take metaphysics seriously.

In this book, we have to a certain extent made metaphysics more mathematical and mathematics more metaphysical. It may appear obvious then that although two historically disparate disciplines have been induced to intersect, at some point their theoretical boundaries should be honored and maintained. At a certain point in the assertions we derive about reality by perceptive application of appropriate transcendental elements, it is inevitable that mathematics and metaphysics will each go their separate ways. However, prior to that systemic divergence, what will the engaged philosopher do to maintain their intersection for as long as the emerging shared elements generate results useful to metaphysical inquiries?

To pursue reality may sound odd, since in a sense we are always within reality—indeed each of us is a part of it. In what sense then is it incumbent on us to "pursue" that which is real? Consider, however, that the word "mathematics" is of Greek origin and, as Liddell-Scott-Jones (the standard lexicon) indicates, is derived from μανθάνω, "to learn." Thus, the adjective μαθημάτικός, "mathematical," would, taken more literally, mean "disposed to learn." To study mathematics is to know about the subject matter of mathematics. But the mathematical person in the original Greek sense is basically someone "disposed to learn." This etymological perspective suggests that what we can learn from mathematics may well range beyond its scope as traditionally formulated and practiced. If so, then mathematics can teach us not only about numbers, lines, figures, equations, functions, integrals, differential equations, group theory, Lie groups, real analysis, Schrödinger equations and beyond, but also about how the import of these classical and modern areas of mathematics—and mathematics in general—can contribute perspectives toward an incisive articulation of other regions of reality. This possibility, if entertained with a measure of seriousness (and sufficient industry), allows the revelatory intersection of mathematics and metaphysics to begin to emerge. The argument of this book has attempted to develop, justify and illustrate this possibility.

Toward the conclusion of *Adventures of Ideas* (294), Alfred North Whitehead cautions that "the speculative methods of metaphysics are dangerous, easily perverted. So is all Adventure; but Adventure belongs to the essence of civilization." Whitehead's notion of Adventure is technical, connected to the ways he viewed civilization as the result of various socializing processes within and between political entities; however, the ordinary, broader sense of the word fits the scope of the approach to metaphysical inquiry developed in this book. Adventure can name the intellectual and philosophical journey, including the expenditure of time, concentrated energy, and imaginative vivification required to work through alternate visions and divergent vocabularies of other metaphysical accounts of reality, then to erect parallel confrontations between (and, surely, among) potentially relevant as well as established metaphysical statements—including, of course, whatever positions the metaphysician, as one person among many, embraces and defends.

If we explore the adventure motif, we appreciate the fact that many adventures proceed hand in hand with risk; if so, then ventures of this sort will often, if not always, require courage. When the popular mind views such phenomena from the perspective of the "ivory tower" syndrome (and, indeed, perhaps also in the standard and typically unconsidered sensibilities of academic philosophers), doing philosophy and the exercise of courage are rarely, if ever, taken as a coordinate pair. Professorial types go off alone and write books (on metaphysics, if they so choose). What in that choice and activity requires courage? But a moment's reflection will make us realize that rubbing shoulders with reality and then realizing, or even just beginning to suspect, that one's metaphysical transcriptions of what is and is not real are not as firm as one had thought will, in most cases, instigate a variety of reactions, some of which are doubtless disquieting, especially for anyone who takes seriously a full-blooded approach to determining the metaphysical dimensions of the philosophical enterprise. The advancement of a metaphysical position in any sort of public forum may be met by more or less restrained criticisms—whether in print (editors' reactions, readers' reports, book reviews), in person (at conferences and meetings), informally (in conversations with colleagues), and perhaps most strikingly, in private moments when the metaphysician is just thinking in solitude, the dialogue of the soul with itself as Plato tersely (and profoundly) says in the *Theaetetus* (see also W. B. Yeats' hauntingly evocative counterpart image, that is, thinking as it occurs in "the deep heart's core").

Metaphysics is a risky pastime then, at least potentially. At this point, a brief and somewhat more concentrated return to Plato is relevant: recall that a few moments before Socrates draws a square in the sand to help educate young Meno in the theoretical subtleties of geometrical demonstration, he asserts to his audience that "the whole of nature is akin" and that it is possible for us to discover everything for ourselves if "we are brave and do not tire of the search." And in another, very different dialogue, Socrates urges Theaetetus to continue with the quest to define knowledge: "never say it is beyond your power; it will not be so, if heaven wills and you take courage" (*Meno*, 81d; *Theaetetus* 151d). Thus, both early and late in the Platonic adventure of producing dialogues aimed at articulating perspectives on truth and reality, we find Socrates emphasizing the need for a fundamental dimension of fortitude when we confront and intend to address difficult theoretical matters.

How much reality do we want to incorporate into our metaphysics? A reply we should probably expect to hear often in our currently splintered philosophical age might be—"Just as much of it as I feel like analyzing—no less but certainly no more." This book has developed, in a reasoned series of speculative steps, a transactional system of broadly based, abstract notions (via equiprimordiality and the transcendental grid) which, if employed with sufficient imaginative energy—and courage—will extend our initial metaphysical dabbling into the company of the great, expansive metaphysical visions of the past and, perhaps, beyond. Furthermore, given that we ourselves are incontrovertibly part of reality, any attempt at producing results in metaphysics is in a very real sense always an exercise in self-scrutiny. It would surely not be idle to wonder whether such examinations could, at least occasionally, lead to measures of progress in the area of "knowing thyself."

Bertrand Russell proclaimed that regarding the highest aspirations of humanity, "the sense of being more," the "touchstone of the highest excellence, is to be found in mathematics, as surely as in poetry." Classic Greek philosophy on number and philosophical method, modern abstract algebra and Abelian groups—this book explores some of what can be said and indeed what can be derived when these expressions of philosophy and mathematics meet and mingle under the sanctioning authority of a uniform foundational reality—which Book VI of Plato's *Republic* imagistically and majestically provides. We may reasonably hope that duly imaginative (and logically attentive) explorations of additional regions of thought shared by the two disciplines will not be sterile exercises. Whether or not any members of a sympathetic audience for the argument of this book can produce something novel and revelatory from this high-level interplay remains to be seen. But the initial effort to try to do so is up to those who take seriously the adventurousness proper to the pursuit of metaphysics. And, in the end, we can determine whether any adventure is successful or a misadventure only after it has been attempted.

Notes

1 Alfred North Whitehead, *Science and the Modern World* (New York: Free Press, 1967), 34.

2 The recent interest in comparative metaphysics suggests that the difference between this area of research and the primary thesis of this book should be

clarified. We are not pitting one metaphysical system against another, whether in whole or in part; rather, the transcendental grid and equiprimordiality offer a range of concepts and preliminary lines of argument for application to a given metaphysical position for purposes of improving that position's overall effectiveness as an account aiming at discursively describing reality or some perspectives on reality. If every attempt at metaphysics were to become as good as it could be, then comparative metaphysics would presumably become even more engaging to those philosophers who have taken the time and effort to maximize control over competing positions in this area of investigation. For an informed review of selected current debates in comparative metaphysics, see: Sophie R. Allen, "Can Metaphysical Structuralism Solve the Plurality Problem?" *International Journal of Philosophical Studies* 26, no. 5 (2018): 722–46. Also, for a recent, incisive, and informed theoretical juxtaposition involving the comparative structure of two disparate ways of being and thinking—the rich immediacy of direct experience and the context-free use of syntactic systems—which is analogous to the main argument of this book in identifying strategies for untangling the intricate "Gordian Knot" of reality, see Timothy E. Eastman, *Untying the Gordian Knot: Process, Reality, Context* (Lanham, MD: Lexington Books, 2022).

3 For a perceptive and penetrating treatment of the relationship between personal and systematic metaphysics, and extensive analyses of different organizational approaches to a wide variety of metaphysical systems, see Stephan Körner, *Metaphysics: Its Structure and Function* (Cambridge: Cambridge University Press, 1984).

4 Bertrand Russell, "The Study of Mathematics," in *Mysticism and Logic and Other Essays* (London: Longman, Green and Co., 1918), 58–73. Quotations from the essay (here, 60) will be cited in the body of the text. The mathematician G. H. Hardy (1877–1947) also emphasized the importance of beauty in mathematics; see his frequently quoted (but not developed in context to the extent that we find in Russell) principle in "A Mathematician's Apology": "The mathematician's patterns, like the painter's and the poet's, must be *beautiful* [italics in text]; the ideas, like the colors and the words, must fit together in a harmonious way." A little later, Hardy comments on the matter of a definition of beauty in this regard: "It may be very hard to *define* [italics in text] mathematical beauty, but that is just as true of beauty of any kind—we may not know quite what we mean by a beautiful poem, but that does not prevent us from recognizing one when we

read it." Russell's narrative concerning mathematical beauty provides points of departure for rendering this kind of beauty less a matter of pure intuition, as it seems to be for Hardy.

5 Aristotle's emphasis on symmetry (σύμμετρία) in the context of the beauty of mathematics found a contemporary response in the thought of the mathematician and metaphysician Hermann Weyl. Weyl (1885–1955) takes as axiomatic that "beauty is bound up with symmetry," a principle that he developed in a number of areas, including science and art. See Hermann Weyl, *Symmetry* (Princeton: Princeton University Press, 1980).

6 David Hume, *An Enquiry Concerning Human Understanding*, 2nd. ed. (Indianapolis: Hackett, 1993), 55.

7 Friedrich Nietzsche, *Beyond Good and Evil*, trans. Walter Kaufmann (New York: Random House, 1966), 3.

8 Just for a moment, we might in this regard enter another form of discourse and heed what the narrator asserts in *Middlemarch* (Book Four, Chapter 34), arguably George Eliot's most philosophical novel: "We are all of us imaginative in some form or other, for images are the brood of desire."

9 René Descartes, *Meditations on First Philosophy*, 3rd ed., trans. Donald A. Cress (Indianapolis: Hackett, 1993), 47.

10 G. W. Leibniz, *Discourse on Metaphysics and Other Essays*, trans. Daniel Garber and Roger Ariew (Indianapolis: Hackett, 1981), 6.

11 Alfred North Whitehead, *Adventures of Ideas* (New York: Macmillan [Mentor Books], 1933), 132.

12 For a useful selection of materials on the notion of productive imagination, see Saulius Geniusas and Dmitri Nikulin, eds., *Productive Imagination: Its History, Meaning and Significance* (Lanham, MD: Rowman & Littlefield, 2016). We may note the likely utility of the productive imagination in extending the applicability of this book's primary line of argument beyond that of group theory, for example, category theory, which represents an instance of the next level of abstraction after that occupied by groups. In category theory (a mid-twentieth-century discovery), mathematical objects, whether groups or other types of mathematical realities, are identified and their relations to one another analyzed and expressed as "morphisms" that are drawn as arrows. Categories are then introduced to study these diagrams of arrows as universal concepts, and the kinds of functions that apply to these relational diagrams. In a metaphysical context, this ascendancy in abstractive content and control mirrors an ongoing ascent toward a kind

of limiting reality, the good as such. Thus, a metaphysician working with the components of category theory could, with due application and precision, hypothetically attain heights of metaphysical theorizing that would, for example, transcend what Plato achieved in positing the realm of the Forms. It should be emphasized, of course, that category theory stands as only one branch of advancement in the history of mathematics. For an informed and comprehensive account of the history of modern mathematics, which, as we have argued in this book, can function as a veritable storehouse of potential metaphysical insight, see, for example, Edna E. Kramer, *The Nature and Growth of Modern Mathematics* (New York: Hawthorn Books, Inc, 1970).

Bibliography

Allen, Reginald E. *Plato's Parmenides*. Revised ed. New Haven: Yale University Press, 1994.

Allen, Sophie R. "Can Metaphysical Structuralism Solve the Plurality Problem?" *International Journal of Philosophical Studies* 26, no. 5 (2018): 722–46.

Aristotle. *Aristotle's Metaphysics*. Edited by W. D. Ross. 2 Vols. Oxford: Clarendon Press, 1924.

Aristotle. *Aristotle's Metaphysics M and N*. Translated by Julia Annas. Oxford: Oxford University Press, 1976.

Aristotle. *The Basic Works of Aristotle*. Edited by Richard McKeon. New York: Random House, 1941.

Bulmer-Thomas, Ivor. "Plato's Theory of Numbers." *Classical Quarterly* 33, no. 2 (1983): 375–84.

Burnyeat, Myles. "Platonism and Mathematics: A Prelude to Discussion." In *Explorations in Ancient and Modern* Philosophy. Online, 213–40. Cambridge University Press, 2012.

Cherniss, Harold. "Plato as Mathematician." *The Review of Metaphysics* 4, no. 9 (1951): 395–425.

Cherniss, Harold. *The Riddle of the Early Academy*. Berkeley: University of California Press, 1945.

Cook Wilson, John. "On the Platonist Doctrine of the ἀσύμβλητοι ἀριθμοί." *The Classical Review* 18 (1904): 247–60.

Cornford, F. M. *Plato's Cosmology*. New York: The Liberal Arts Press, 1957.

Descartes, René. *Meditations on First Philosophy*. 3rd ed. Translated by Donald A. Cress. Indianapolis: Hackett, 1993.

Eastman, Timothy E. *Untying the Gordian Knot: Process, Reality, Context*. Lanham, MD: Lexington Books, 2022.

Florin, George Callan. "Numbers, Ontologically Speaking: Plato on Numerosity." In *Numbers and Numeracy in the Greek Polis,* edited by Robert Sing, 219–36. Leiden: Brill, 2021.

Geniusas, Saulius and Dmitri Nikulin, eds. *Productive Imagination: Its History, Meaning and Significance.* Lanham, MD: Rowman & Littlefield International, 2016.

Graeser, A., ed. *Mathematics and Metaphysics in Aristotle.* Bern: Paul Hapt, 1987.

Halper, Edward C. *One and Many in Aristotle's Metaphysics.* Las Vegas: Parmenides Publishing, 2009.

Heath, Sir Thomas. *A History of Greek Mathematics.* Vol. 1. New York: Dover, 1981.

Hume, David. *An Enquiry Concerning Human Understanding.* 2nd ed. Indianapolis: Hackett, 1993.

Klein, Jacob. *Greek Mathematical Thought and the Origin of Algebra.* Translated by Eva Braun. Cambridge, MA: The M.I.T. Press, 1968.

Körner, Stephan. *Metaphysics: Its Structure and Function.* Cambridge: Cambridge University Press, 1984.

Kramer, Edna E. *The Nature and Growth of Modern Mathematics.* New York: Hawthorn Books, Inc., 1970.

Lear, Jonathan. "Aristotle's Philosophy of Mathematics." *The Philosophical Review* 91, no. 2 (1982): 161–92.

Leibniz, G. W. *Discourse on Metaphysics and Other Essays.* Translated by Daniel Garber and Roger Ariew. Indianapolis: Hackett, 1981.

Mohr, Richard D. "The Number Theory in Plato's *Republic VII* and *Philebus.*" *Isis* 72, no. 4 (1981): 620–7.

Nietzsche, Friedrich. *Beyond Good and Evil.* Translated by Walter Kaufmann. New York: Random House, 1966.

Plato. *The Collected Dialogues of Plato.* Edited by Edith Hamilton and Huntington Cairns. New York: Pantheon Books, 1961.

Pritchard, P. *Plato's Philosophy of Mathematics.* Sankt Augustin: Academia Verlag, 1995.

Roochnik, David. "Counting on Number: Plato on the Goodness of *Arithmos.*" *American Journal of Philology* 115 (1994): 543–63.

Ross, Sir David. *The Right and the Good.* Oxford: at the Clarendon Press. 1930.

Russell, Bertrand. "The Study of Mathematics." In *Mysticism and Logic and Other Essays*, 58–73. London: Longman, Greens and Co., 1918.

Taylor, A. E. "Forms and Numbers: A Study in Platonic Metaphysics." *Mind* 35 (1926): 419–40.

Taylor, A. E. *Plato The Man and His Work.* New York. Meridian Books, 1956.

Wedberg, Anders. *Plato's Philosophy of Mathematics.* Stockholm: Almquiste Wyksell, 1955.

Weyl, Hermann. *Symmetry.* Princeton: Princeton University Press, 1980.

White, David A. *Myth and Metaphysics in Plato's* Phaedo. Selinsgrove: Susquehanna University Press, 1989.

White, David A. *Myth, Metaphysics and Dialectic in Plato's* Statesman. Aldershott: Ashgate [Routledge, 2016], 2007.

Whitehead, Alfred North. *Adventures of Ideas*. New York: Macmillan [Mentor Books], 1933.

Whitehead, Alfred North. "Mathematics and the Good." In *The Philosophy of A. N. Whitehead*. 2nd ed., 666–81. LaSalle, IL: Open Court, 1999.

Whitehead, Alfred North. *Modes of Thought*. New York: Free Press, 1968.

Whitehead, Alfred North. *Process and Reality*. Corrected ed. New York: Free Press, 1978.

Whitehead, Alfred North. *Science and the Modern World*. New York: Free Press, 1967.

Whitehead, Alfred North and Bertrand Russell. *Principia Mathematica To *56*. Cambridge: At the University Press, 1962.

Index

Note: Page numbers followed by 'n.' represent notes.

www.ingramcontent.com/pod-product-compliance
Ingram Content Group UK Ltd.
Pitfield, Milton Keynes, MK11 3LW, UK
UKHW020213290926
15470UKWH00001B/155